Practical Statistics for Environmental and Biological Scientists

Practical Statistics for Environmental and Biological Scientists

John Townend

University of Aberdeen, UK

JOHN WILEY & SONS, LTD

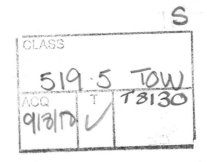

Copyright © 2002   John Wiley & Sons Ltd, The Atrium, Southern Gate, Chichester,
West Sussex PO19 8SQ, England

Telephone   (+44) 1243 779777

Email (for orders and customer service enquiries): cs-books@wiley.co.uk
Visit our Home Page on www.wileyeurope.com or www.wiley.co.uk

Reprinted with corrections March 2003, July 2004, April 2005

Other Wiley Editorial Offices

John Wiley & Sons Inc., 111 River Street, Hoboken, NJ 07030, USA

Jossey-Bass, 989 Market Street, San Francisco, CA 94103-1741, USA

Wiley-VCH Verlag GmbH, Boschstr. 12, D-69469 Weinheim, Germany

John Wiley & Sons Australia Ltd, 33 Park Road, Milton, Queensland 4064, Australia

John Wiley & Sons (Asia) Pte Ltd, 2 Clementi Loop #02-01, Jin Xing Distripark, Singapore
129809

John Wiley & Sons Canada Ltd, 22 Worcester Road, Etobicoke, Ontario, Canada M9W 1L1

British Library Cataloguing in Publication Data

A catalogue record for this book is available from the British Library

ISBN 10: 0-471-49664-2 (HB)   ISBN 13: 978-0-471-49664-9 (HB)
ISBN 10: 0-471-49665-0 (PB)   ISBN 13: 978-0-471-49665-6 (PB)

Typeset in 10.5 / 13pt Times by Vision Typesetting Manchester
Printed and bound in Great Britain by TJ International, Padstow, Cornwall
This book is printed on acid-free paper responsibly manufactured from sustainable forestry
in which at least two trees are planted for each one used for paper production.

Contents

Preface

Statistics wasn't forced upon the environmental and biological sciences; it has been absorbed into their practice because it was realized that it had something to offer. Statistical methods provide us with ways of summarizing our data, objective methods to decide how much confidence we can place in experimental results, and ways of uncovering patterns that are initially masked by the complexity of a dataset. Also, if we carry out scientific investigations according to our instincts, there is a risk that we will bias the results by overlooking some important factor or through our desire to get a particular result. By carefully following accepted statistical procedures we can avoid these problems and, just as importantly, we will be seen to have avoided them, so our results will be more readily accepted by others.

Statistics is also a useful means of communication. For example, a researcher might state that 'the molluscs had a mean shell length of 12.2 mm $\pm$ 1.6 mm (standard error)', or report that 'ANOVA showed significant differences between nitrogen contents in different groups of plants ($P = 0.02$)'. These are succinct ways of explaining a great deal of detail about how studies have been carried out and what can be concluded from them. Of course, they are only really a useful means of communication if you understand what the terms mean. Like it or not, though, they are widely used, so whether you intend to use statistics yourself or just read about others' research, it will still be a great help to know something about it.

While teaching statistics in a university I found that, for the most part, the statistical methods required by both environmental and biological scientists were the same. Indeed this might be expected, because much of the science is common to both as well. I also found that requirements were very similar at all levels from undergraduate to experienced professional. Really there is seldom any necessity to use complex statistical methods to do world-class research in environmental and biological sciences. Those who are able to identify the key, simple questions to ask are likely to enjoy the greatest success. So it is that I have tried to put together a book that addresses as many of the most common needs as possible.

The choice of content is based on the questions I have most frequently been asked and the explanations that seemed to work best. Memorizing formulae will be of very little practical use to you, except perhaps to pass an exam; most calculations can be carried out by computer these days. However, computers do not generally tell you whether you are carrying out the right calculations or exactly what you can conclude from the results. Here textbooks still have a part to play. In this book I try to unlock many of the codes commonly used to present scientific information and to provide you with the tools you need to be an effective user of statistics yourself. I wholeheartedly hope that it will provide you with the information you need.

PART I

STATISTICS BASICS

Chapters 1 to 6 introduce the ideas behind statistical methods and how practical studies should be set up to use them. They aim to give the required background for using the methods in Part II. Readers who are new to statistics or in need of a short refresher might find it useful to read this part in its entirety.

1

Introduction

If your first love was statistics, you probably wouldn't be studying or working in environmental or biological sciences. I am starting from this premise.

1.1 Do you need statistics?

Somebody who is trying to sell you a statistics textbook is probably not the best person to ask whether you need statistics. Maybe you have opened this book because you have an immediate need for these techniques or because you have to study the subject as part of a course. In this case the answer for you is clearly yes, you need statistics. Otherwise, if you want to know whether statistics is really relevant to you, ask people who have been successful in your chosen area – academics, researchers or people doing the kind of job you want to do in the future.

Some use it more than others, and certainly you will find some very successful people who are not confident with statistics and possibly dislike any involvement with it. I don't believe being a brilliant statistician is a necessary condition for being a brilliant biologist or environmental scientist. However, you will probably find that most of the people you ask would have found it useful to understand statistics at some stage in their career, perhaps very regularly. Even if you do not need it to present results yourself, you will need to understand some statistics in order to understand the real meaning of almost any scientific information given to you.

The fact that most university degrees in environmental and biological sciences include a compulsory statistics course is simply a recognition of this. However, do not think that understanding statistics is all or nothing. Even a basic understanding of why and when it is used will be very valuable. If you can grasp the detail too, so much the better.

1.2   What is statistics?

Football scores, unemployment rates and lengths of hospital waiting lists are statistics, but not what we commonly think of as being included in the subject of statistics. An interesting definition I heard recently was that statistics is 'that part of a degree which causes a sinking feeling in your stomach'. I don't have an all-encompassing definition myself, but it will be helpful if you can keep in mind that more or less everything in this book is concerned with trying to draw conclusions about very large groups of individuals (animate or inanimate) when we can only study small samples of them. The fact that we have to draw conclusions about large groups by studying only small samples is the main reason that we use statistics in environmental and biological science.

Supposing we select a small sample of individuals on which to carry out a study. The questions we are trying to answer usually boil down to these two:

- If I assume that the sample of individuals I have studied is representative of the group they come from, what can I tell about the group as a whole?

- How confident can I be that the sample of individuals I have studied was like the group as a whole?

These questions are central to the kind of statistical methods described in this book and to most of those commonly used in practical environmental or biological science. We are usually interested in a very large group of individuals (e.g. bacteria in soil, ozone concentrations in the air at some location which change moment by moment, or the yield of wheat plants given a particular fertilizer treatment) but limited to studying a small number of them because of time or resources.

Fortunately, if we select a sample of individuals in an appropriate way and study them, we *can* usually get a very good idea about the rest of the group. In fact, using small, representative samples is an excellent way to study large groups and is the basis of most scientific research. Once we have collected our data, our *best estimate* always has to be that the group as a whole was just like the sample we studied; what other option do we have? But in any scientific study, we cannot just assume this has to be correct, we also need to use our data to say how confident we can be that this is true. This is where statistics usually comes in.

Almost all experimental results are as described above. They state what is the case in a small sample that was studied, and *how likely* it is to be true of the group it was taken from. Elementary textbooks often quote results leaving out any indication of how much confidence we can place in them for the sake of clarity. However, most of the results they quote originally come from papers published in scientific journals. If you look at the results presented in a scientific journal, you will see statements like:

Big gnomes catch more fish than little gnomes ($P = 0.04$)

The study would have been carried out using *samples* of big gnomes and small gnomes and the statement is really shorthand for:

- In our samples, on average, big gnomes caught more fish than little gnomes, so we expect that big gnomes in general catch more fish than little gnomes.

- Based on the evidence of our samples, we can really only be 96% confident that big gnomes in general do catch more fish than little gnomes.

You can see that the second, qualifying, statement (which comes from the $P = 0.04$) is really quite important to understanding what the researchers have actually shown. It is not as clear-cut proof as you might otherwise think.

We will look in more detail at how to interpret the various forms of shorthand as we go through the different statistical techniques, but notice that when the result is stated in full we have (i) a result for the whole group of interest *assuming* that the samples studied were representative, and (ii) a measure of *confidence* that the samples studied actually were representative of the rest of the groups. This point is easy to lose sight of when we start to look at different techniques.

Textbooks tend to emphasize *differences* between statistical techniques so that you can see when to use each. However, these same ideas lie behind nearly all of them. Statistical methods, in a wide variety of disguises, aim to quantify both the effects we are studying (i.e. what the samples showed), and the confidence we can have that what we observed in our samples would also hold for the rest of the groups they were taken from. If you can keep this fact in mind, you already understand the most important point you need to know about statistics.

1.3 Some important lessons I have learnt

Statistics as a science in its own right can be very complicated. The statistics you need to be a good environmental scientist or biologist is only a small and fairly straightforward subset of this. Even a general understanding of the basic ideas will be a great asset when you come to interpret other people's experimental results. When you know some of the shorthand, like the example of the gnomes, you will see that very many scientific 'facts' are not as clear-cut and certain as we often imagine. Understanding just this already gives you statistical and scientific skills beyond those of the general public. You will quickly learn to be more discerning about what scientific 'facts' you really believe.

There is no denying that a skilled statistician would have methods in his or her armoury beyond those I have included in this book. There are not statistical techniques available for *every* eventuality, but there are techniques for a good many of them. However, it takes rather a long time to learn about them all and probably you want to get on with some environmental or biological science too. I have therefore selected in this book a range of techniques that I consider most relevant and useful, and I believe these are sufficient to allow you to conduct most types of environmental or biological study *with a little careful planning*. Now here's the bit that a lot of people find difficult to grasp. The thing that separates competent environmental scientists and biologists from incompetent ones, in terms of statistical skills, is not numeracy, but *careful planning*. The chances are that a computer will do all of your calculations for you.

By the time you sit down at the keyboard with your data you will have already made most of the mistakes you are likely to make. Just when you think you are about to *start* the statistical part of your project, your part in the statistics is really coming to an end. If you have planned carefully, formed a clear idea of what you are investigating, followed the layout of appropriate examples from this or other books, and carried out your survey or experiment accordingly, the analysis and interpretation will be plain sailing. Please don't leave all thought of statistical analysis to the point where you sit down with your data already in hand. You would be unlikely to find the analysis plain sailing then. This is an important lesson I have learnt.

1.4   Statistics is getting easier

Until the 1980s most statistical calculations were done using a pocket calculator or by hand. Nowadays almost all calculations are carried out by computer. We need only know which test to use and how to enter the data in order to carry it out. I have heard concerns that many students nowadays just quote the output without understanding it. This is probably true, but it was always thus. As far as I can see, the only difference with precomputer days is that then you would spend two hours struggling with the calculations so there was a feeling you had earned the right to give the result. I don't believe the average user of statistics either knew or cared what the calculations were actually doing any more then than they do now.

Although I do not think that as *users* of statistics we need to do the calculations ourselves, we do lose a lot if we take the results without understanding anything about the methods. Until recently it was *necessary* to teach the calculations behind statistics because without them you could not use statistics, whether you understood them or not. To someone who is comfortable with mathematical concepts, the formulae are also a satisfactory explanation of what is going on, so teachers often believed they had covered method and understanding at the same time.

An aunt of mine used to say, 'There are liars, damn liars and sadistics. Most of the liars and damn liars go into law or advertising so don't bother us much, but most of the sadistics teach numeracy skills. That's why maths and statistics are hard.' It is my belief that because statistics has traditionally been taught as a mathematical skill, although most students got by with the methods, very few picked up the understanding along the way. There is a great challenge here for teachers of statistics. Rather than seeing the removal of the calculations as a sad loss to understanding, we should take advantage of this to try to make the meaning and value of statistics more accessible to all.

1.5 Integrity in statistics

Scientific research *relies* on the integrity of the people conducting the research. Most of the time, we just have to believe that researchers have been honest in their work as there is no way to tell if results have been made up. In fact, in my experience very few people do lie about the actual values they have collected, even if they are disappointing. Most scientists, I think, have a fairly strong sense of conscience. We also need to have this attitude to carrying out an appropriate statistical analysis. Some kinds of analysis are easier to do than others and some may appear to give us the result we want whilst others do not. However, just because it is *possible* to use one statistical technique does not necessarily mean it is valid. Usually it is necessary to make certain checks on the data to discover whether a particular method can be applied validly (Chapter 6). Unfortunately, this can sometimes lead us to have to do more work, so there is a temptation to skip this stage.

The reader of our work, of course, has to assume that we *have* done the appropriate checks and, if necessary, carried out the extra work. Otherwise we should add the qualifying statement 'assuming the test was valid in this case', but then who would take our results seriously? If we just present results without checking the validity of using our chosen statistical method, we are deliberately deceiving the reader. If you value the integrity of your work, therefore, checking the validity of applying particular statistical methods must be seen as part of the normal process of statistical analysis. The checks required are described in the 'Limitations and assumptions' sections preceding each of the methods described in this book.

1.6   About this book

I have tried as far as possible to avoid mathematical descriptions of the techniques. There are a few simple formulae which readers might find occasion to use because they are not covered by some of the common statistical programs. I have included these in boxes; you can skip them if you want. I have also included some formulae in Appendix A, principally because they might be needed by some people for examinations in statistics. Mainly I have tried here to describe some of the range of techniques available, when you can use them, how to use them, and what the results are telling you; I have assumed that you will use a computer to do most of the calculations.

Competence and confidence in statistics will be an asset to you as an environmental or biological scientist, but at the same time it is only one of many things that will make you a good environmental or biological scientist. You only have so much time available, and to suggest you study the detail of statistics may not be the best use of it. With this in mind I have tried to include only techniques and a level of detail that I think will be genuinely useful to those studying or working in environmental or biological sciences.

There are different schools of thought about whether or not one should illustrate statistics with real experimental data. My own thoughts on this are that it is best to use simple datasets to demonstrate the techniques. It is not possible to cover all the eventualities that will arise in real-life results. However, provided you understand clearly what is required, you will be in a strong position to decide how to collect and handle your own data. All of the datasets in this book are therefore invented to demonstrate particular points.

The book is divided into two parts. Chapters 1 to 6 cover some basic statistical ideas and are intended to give you the necessary background for any of the statistical techniques in later chapters. Chapters 7 to 15 are more of a reference section with different statistical tests or methods described in each chapter. Guidance on which test to use in a particular situation is given in Section 3.2 and in the decision chart in Appendix D.

I have also included some pointers to more advanced techniques that readers might find useful in the further reading sections at the end of some of the chapters. If you have a computer package available to carry these out, understanding the details of the calculations need not necessarily be a problem to you. Nevertheless, before going ahead and using any of them it is important to familiarize yourself with what the tests are actually testing, and the assumptions and limitations they have about the types of data they are suitable for. In general, I have not specified particular texts to consult because these techniques are widely covered in many of the more in-depth statistical textbooks, and probably most of these would give you similar information.

2

A Brief Tutorial on Statistics

2.1 Introduction

From Chapter 3 onwards I describe a range of statistical tests and methods, and how to design experiments or surveys that make use of them. If you are studying this subject for the first time, you will probably find it difficult to retain all this information in your head. For the most part, this is not a problem. You can refer back to the book when you need to. However, there are some basic ideas behind all statistical methods and if you can keep these in mind, they will help you to make sense of statistical methods in general. These basic ideas are the subject of this chapter.

2.2 Variability

Think of a group you might want to study, e.g. the lengths of fish in a large lake. If all of these fish were the same length, you would only need to measure one. You can probably accept that they are not all the same length, just as people are not all the same height, not all volcanic lava flows are the same temperature, and not all carrots have the same sugar content. In fact, most characteristics we might want to study vary between individuals. If we measured the lengths of 100 fish, we could plot them on a graph as in Figure 2.1(a).

To understand this graph, think how we would add the extra point if we measured another fish to be 42 cm. It would appear as an additional fish in the column labelled > 40–45 cm. The graph tells us that most of the fish were about the same length, and gives us a picture of how widely spread the individuals' lengths were around this. We can see that only a few fish were greater than 50 cm or less than 15 cm. Figure 2.1(b) shows the more usual ways of presenting this kind of data.

(a)   (b)

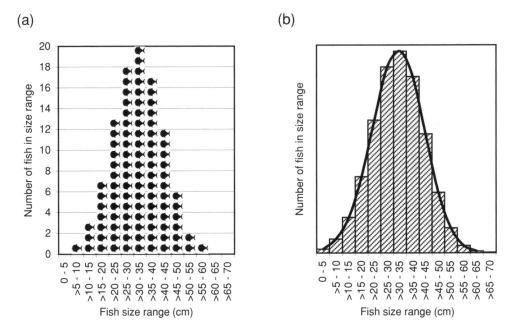

Figure 2.1 (a) Numbers of fish out of a sample of 100 falling into different size ranges. Note >10–15 means fish more than 10 cm long, up to and including 15 cm. (b) When larger numbers of measurements are involved, it becomes inconvenient to represent each individual, so a column graph or a line can be used to show the shape of the distribution

The graphs in Figure 2.1 are called *frequency distributions*. For some things we might measure, we would find different distributions such as a lot of low values, some high values, and a few extremely high values. These result in different shapes of graph (Sections 6.1 and 6.2). However, it turns out that if we measure a set of naturally occurring lengths, concentrations, times, temperatures, or whatever, and plot their distribution, very often we do get a diagram with a shape similar to those in Figure 2.1. This shape is called a *Normal distribution*. Statisticians have derived a mathematical formula which, when plotted on a graph, has the same shape. Being able to describe the distribution of individual measurements using a mathematical formula turns out to be very useful because, from only a few actual measurements, we can estimate what other members of the population are likely to be like. This idea is the basis of many statistical methods.

2.3   Samples and populations

As discussed in Chapter 1, practical considerations almost always dictate that we study any group we are interested in by making measurements or observations on a relatively small *sample* of individuals. We call the group we are actually interested in the *population*. A population in the statistical sense is fairly close to the common meaning of the word, but can refer to things other than people, and usually to some particular characteristic of them. Here are some examples of statistical populations:

- The lengths of blue whales in the Arctic Ocean

- All momentary light intensities at some point in a forest

- Root lengths of rice plants of a particular variety grown under a specific set of conditions

In the first example, the population is real but we are unlikely to be able to study all of the whales in practice. Populations in the statistical sense, however, need not be finite, or even exist in real life. In the second example, the light intensity could be measured at any moment, but the number of moments is infinite, so we could never obtain measurements at every moment. In the third example, the population is just conceptual. We really want to know about how rice plants of this variety *in general* would grow under these conditions but we would have to infer this by growing a limited number of rice plants under the specified conditions. Although the few plants in our sample may be the only rice plants ever to be grown in these conditions, we still consider them to be a sample representing rice plants of this variety in general growing in these conditions.

2.4 Summary statistics

It is often useful to be able to characterize a population in terms of a few well-chosen statistics. These allow us to summarize possibly large numbers of measurements in order to present results and also to compare populations with one another.

Mean, variance, standard deviation and coefficient of variation

If we want to describe a population it may sometimes be useful to present a frequency distribution like those in Figure 2.1, but this is usually more information than is needed. Two items are often sufficient:

- A measure which tells us what a 'typical' member of the population is like

- A measure which tells us about how spread out the other members of the population are around this 'typical' member

To represent a 'typical' member of the population, we usually use the *mean* (all

Box 2.1   Standard deviation and variance

One way to characterize how spread out the values in a sample are would be to calculate the difference between each measurement and the sample mean, and then to calculate the mean of these differences.

Here's an example:

Sample value		8	12	10	7	7	11	8	mean = 9.0
Difference from mean		1	3	1	2	2	2	1	mean of differences = 12/7 = 1.7

Had statistics been developed after computers became readily available, this might be the measure of spread we commonly use; indeed some recently developed statistical methods do use this. However, when calculations were done by hand, it was found to be more convenient to use the mean of the *squares* of the differences as a measure of spread; there are mathematical shortcuts to getting the result in this case which are useful when you have a large sample. Since a great deal of statistical theory and tests built up around this, we still use it today.

For the above sample:

Sample value	8	12	10	7	7	11	8	mean = 9.0
Difference from mean	1	3	1	2	2	2	1	mean of differences = 12/7 = 1.7
Square of difference from mean	1	9	1	4	4	4	1	mean of squared differences = 24/7 = 3.4

When we take a random sample it may or may not include the largest and smallest values in the population, yet these would both contribute the largest *squares* of differences from the mean. Since they are not present in all samples, *on average* the mean of the squares of differences is less when it is calculated from a sample than if it was calculated for the population as a whole.

However, what we really want is to estimate the spread of values in the population, not in the sample itself, so we need to correct for this. This is done by modifying the above calculation so that we divide not by the number of values in our sample, but by *one less than the number of values in our sample*. You can see from the example below that this corrects things in the right direction. Some complex statistical theory shows that this simple modification corrects the value by the right amount.

For the above sample again:

Sample value	8	12	10	7	7	11	8	mean = 9.0
Difference from mean	1	3	1	2	2	2	1	mean of differences = 12/7 = 1.7
Square of difference from mean	1	9	1	4	4	4	1	corrected mean = 24/6 = 4.0

This figure – the corrected mean of the squared differences – is called the *variance* and is an unbiased *estimate* of the spread of values in the population, calculated from a sample.

Variance has units, e.g. if the measurements had been in grams (g), the variance would be in units of square grams (g^2). The square root of the variance is called the *standard deviation*. The standard deviation in the above example is $\sqrt{4.0} = 2.0$, i.e. standard deviation is an alternative measure of the spread of values. Standard deviation has the same units as the actual measurements, e.g. if the measurements had been in grams, the standard deviation would also be in grams. The mathematical formulae for variance and standard deviation are given in Appendix A.

of our values added together then divided by the number of values). In common usage people often refer to this as the 'average' but the term 'mean' is preferred in technical writing.

To express how spread out the individual values in a population are, we usually use the *standard deviation* or *variance*; variance is simply the standard deviation squared (Box 2.1). In describing a population, we might therefore say, 'The mean length of fish in the lake was 32 cm with a standard deviation of 10 cm.' This tells us that most (approximately 68%) of the fish had lengths in the range 22–42 cm. Popular texts and the media often just give the mean with no measure of spread, but as scientists we should recognize that both measures are important. In a different lake the fish might have the same mean length but a very different spread of values. This might have important scientific implications. For example, if big fish eat little fish, the ecology of a lake with a wide range of sizes may be very different to that in a lake where the fish are all about the same size.

A further measure sometimes used to characterize the variability of a group is the *coefficient of variation* (CV). If we are told that the standard deviation of the lengths of ants is 3 mm, and the standard deviation of the lengths of dogs is 20 cm, we could correctly interpret this to mean that dogs are much more variable in their lengths. But we might also want to know which is more variable in relation to its size. If we divide the standard deviation by the mean, we get the coefficient of variation, i.e. the CV is the standard deviation relative to the mean size of the individuals. For convenience, let's suppose the mean length of ants is 10 mm and the mean length of dogs is 100 cm. The CVs are therefore as follows:

- For the ants $3/10 = 0.3 = 30\%$

- For the dogs $20/100 = 0.2 = 20\%$

Relative to their size, ants are more variable. The mean and standard deviation or CV are useful statistics to use to present the results of a survey.

Standard error and 95% confidence interval

In experimental and survey work we are rarely interested in the samples we have actually studied. Our real interest is in the *populations* they come from. This is important to keep in mind, otherwise statistical tests make no sense.

Figure 2.2 shows the soil temperatures at 20 mm depth at different points in a field (the population). Suppose we want to know the overall mean temperature at 20 mm depth in this field – the *population mean*. The only way to find this out for sure would be to measure at every point, which would be impractical in a real field. What should we do? In most cases the best we can do it to use a *sample* of points and study them. We might make a start by measuring at 10 randomly

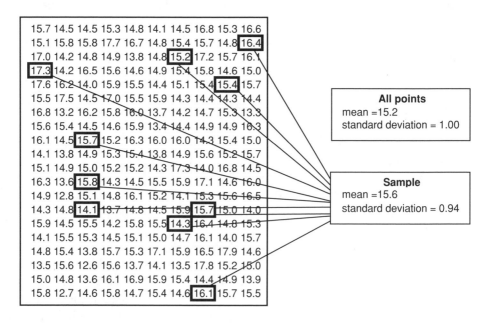

Figure 2.2   Temperatures (°C) at 20 mm depth at 200 points in a field, and the mean and standard deviation of all the points and of a random sample of 10 points

selected points and calculating their mean temperature, the *sample mean* – we just add up the measurements and divide by 10 (Figure 2.2).

Of course, this still doesn't actually tell us the mean temperature for the whole field, it just tells us the mean temperature for those particular 10 points. Why is that any use? Figure 2.3(a) shows the distribution of temperatures in our sample, together with the distribution of temperatures for all the points given in Figure 2.2. As in this case, the distribution of values in a randomly selected sample is usually similar to that in the population as a whole, just with a lot fewer values. We can therefore say that a sample mean is probably a reasonable *estimate* of the population mean. What we need to do now is to try to give some measure of the 'margin of error' in this, i.e. to say, in any particular case, how reliable this estimate is likely to be.

So how can we tell what the margin of error is? Suppose I measured the temperature at another 10 points in the field at random and calculated their mean. I would probably get a slightly different value from the first sample. I could repeat this any number of times. Figure 2.3(b) shows the distribution of a series of 50 *sample means* obtained in this way. To understand this graph, imagine you took another random sample of 10 points and found their mean value was 15.6°C. This new point would be added to the top of the column labelled > 15.5–16.

Notice in Figure 2.3(b) that although the different samples did not all give the

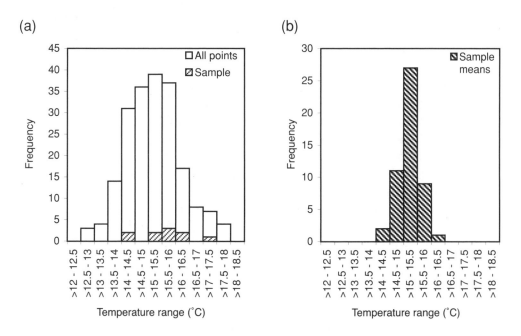

Figure 2.3   (a) Distributions of the temperatures in the field and of the measurements in the sample. (b) Distribution of mean values from 50 random samples, each of 10 measurements

same mean, they are all quite close to one another, and are clustered around the population mean of 15.2°C. Sample means always cluster round the population mean like this. In this particular case, we can see that most of the sample means were within the range 14.5 to 16.0 (i.e. within 0.8 of the population mean). In other words, if we measured and calculated the mean of a randomly selected sample of 10 measurements from this field, we could be reasonably confident that the mean for the whole field would be within the range ±0.8 of it. We have a measure of the 'margin of error' in this estimate, right here.

Although successive sample means are always clustered round the population mean like this, in other situations they may be clustered more or less tightly. To give a 'margin of error' round any particular sample mean, we need to know how tightly clustered a series of *similar* sample means would be in that situation. We could find this out by repeating every experiment 50 or so times but this would be impractical. However, it turns out that we can also estimate how tightly clustered a series of similar sample means would be from a single sample, by applying a simple formula (Box 2.2).

Looking at a graph like Figure 2.3(b) and saying that most means seem to be in a particular range is a bit subjective. Using the formulae in Box 2.2 to calculate the range of values in which most sample means would lie instead, achieves the same thing but also has the advantage that we get a more clearly defined and objective measure of the 'margin of error'. The most commonly used measure is called the *standard error* (Box 2.2). This is the range of values in which we can be approximately 68% confident that the true population mean lies. None of these 'margins of errors' *absolutely* defines the margin of error; the population mean might be a lot further out than this; we will never know for sure.

Therefore if we want to quote a 'margin of error' in which we can be more confident that the population mean really lies (in fact 95% confident), we can use the 95% *confidence interval* (95% CI). The 95% CI gives approximately twice as wide a range of values as the standard error (Box 2.2).

The result of a study may be stated as 12.5 mg ± 1.2 mg. This tells us that the mean value for the *sample* studied was 12.5 mg so this is our best estimate for the population mean. It also tells us that there is a considerable 'margin of error' in this estimate but we could at least be reasonably confident that the population mean was really somewhere in the range 11.3 to 13.7 mg. Either the standard error or the 95% CI can be shown in this way, so the text should make it clear which is given in any particular case.

A lot of people confuse the terms 'standard deviation' and 'standard error', presumably because they are both introduced about the same time in a course, both are new concepts, and both contain the word 'standard'. Concentrate on the words 'deviation' and 'error':

Box 2.2   Standard error and 95% confidence interval

If we repeatedly measure randomly selected samples of, say, 5 individuals from a population, each time the sample mean will probably be slightly different. However, we would find that usually the sample mean was somewhere close to the mean of the population it came from. In fact, if we repeatedly measured samples of, say, 10 individuals, we would find that the means of these samples were even more closely clustered around the population mean (Figure 1). Remember how to interpret frequency distributions: the values are shown on the horizontal axis and the higher the line, the more often that value occurs.

We can calculate the standard deviation, i.e. the spread, of the *individuals* in the population as described in Box 2.1. It is also possible to calculate the standard deviation of the *sample means*. Figure 1 shows that the standard deviation of sample means is less for samples of size 10 than for samples of size 5. The standard deviation of the sample means is called the *standard error* (s.e.).

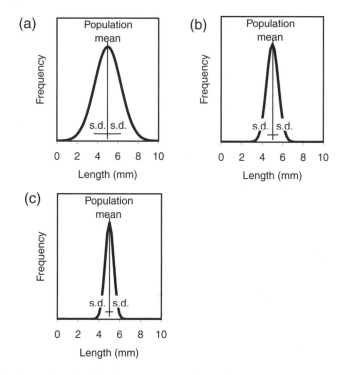

Figure 1   Frequency distributions. (a) For a population. (b) For the means of a series of randomly selected samples of 5 individuals taken from the population in (a). (c) For the means of a series of randomly selected samples of 10 individuals taken from the population in (a). The standard deviation (s.d.) for each of the distributions is also shown

If we know the standard error, we can say that most randomly selected samples would have means in the range

$$\text{population mean} \pm \text{standard error}$$

Perhaps more usefully, if we know the mean of a particular sample, we can be fairly confident that the *population* mean lies in the range

$$\text{sample mean} \pm \text{standard error}$$

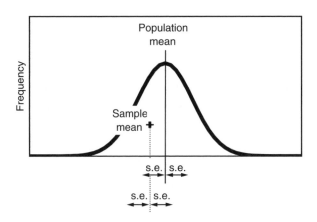

Figure 2   Frequency distribution for the means of a series of similar random samples taken from a population. Most means lie within one standard error (s.e.) of the population mean. Also, given a particular sample mean, in most cases the population mean will lie within one standard error of it

However, in a practical experiment or survey, we usually only take one sample of individuals from a population, so we cannot calculate the standard deviation of the sample means using the same formula that we would use to calculate the standard deviation of a set of individual measurements. Instead we use the relationship

$$\text{standard error} = \frac{\text{standard deviation of the individuals in the sample}}{\sqrt{\text{sample size}}}$$

For example:

Sample values	8, 12, 10, 7, 7, 11, 8	mean $= 9.0$
		standard deviation $= 2.0$
		sample size $= 7$
		standard error $= 2.0/\sqrt{7}$
		$= 0.76$

The standard error gives the range in which we can be approximately 68% confident that the population mean really lies. In this case there is approximately a 68% chance that the true population mean lies in the range 9.0 ± 0.76, i.e. between 8.24 and 9.76. Even if the population mean is really outside this range, it will probably not be far outside.

It is possible to calculate ranges for other degrees of certainty. People sometimes quote the range in which we can be 95% confident that the true population mean lies. This is called the 95% *confidence interval*:

$$95\% \text{ CI} = \text{standard error} \times t_{df,0.05}$$

The value $t_{df,0.05}$ is called a *t*-value and must be obtained from *t*-tables or a computer. The tables have rows for different numbers of degrees of freedom (df). When using the formula above, the rule for calculating the df is

$$df = \text{sample size} - 1$$

In the tables, read down the column headed 0.05 to the correct row for the number of degrees of freedom; if a table states that it gives one-tail values, use the column headed 0.025. You will be looking in the correct column if the number at the bottom of the column, i.e. for $df = \infty$ is 1.96.

For the above sample,

$$95\% \text{ CI} = \text{standard error} \times t_{6,0.05} = 0.76 \times 2.447 = 1.85$$

Therefore we can say that there is a 95% chance that the true population mean lies in the range 9.0 ± 1.85, i.e. between 7.15 and 10.85. The mathematical formulae for standard error and 95% CI are given in Appendix A.

- Standard deviation is a measure of how much *deviation* there is between individuals in a population.

- Standard error is a measure of the 'margin of *error*' involved in estimating the mean of a population.

2.5   The basis of statistical tests

Most statistical tests are about looking for differences between different populations. There is another group concerned with looking at whether different measurements (e.g. height and weight) are related. In all cases we get a result in two parts:

• Whether the populations are different (or measurements are related) if we *assume* our samples are like the rest of the populations.

• How confident we can be that our samples are like the rest of the populations.

As we saw in Section 2.4, successive randomly selected samples generally produce slightly different means, even when they are drawn from the same population. So the samples we are comparing probably *will* suggest one or other of the populations has a greater mean value. Consequently, the main interest in a statistical test is in how *confident* we can be that any differences which appear between the samples are also representative of the populations they came from.

Suppose we want to compare the mean photosynthesis rates of birch seedlings growing at 10°C with ones growing at 20°C. We might grow samples of 15 plants at each temperature to test this. Very likely we will get two different sample means, which leaves us with two possibilities:

• The two populations differ.

• The two populations are the same (i.e. temperature has no effect) but we got two different sample means because of variability between individuals.

Not a very helpful experiment you might think! However, we saw in Section 2.4 that from a sample we could calculate the sample mean and a 'margin of error', a *range* of values where we can be fairly confident that the population mean really does lie. If two samples suggest very different ranges of likely values for the means of the populations they came from, it would be reasonable to conclude that the populations probably were different. However, if there is a large degree of overlap in these ranges, we cannot be confident that the samples really came from different populations. This is essentially the approach taken in statistical testing (Figure 2.4).

Figure 2.4(d) shows that it is not always easy to decide from a graphical presentation of the data whether the populations that two samples came from probably were identical or not. Statistical tests are used to give us objective assessments of this. Typically we will get a result like:

Mean photosynthesis rates were greater in the seedlings growing at 20°C than in those growing at 10°C ($P = 0.016$).

This tells us that the mean photosynthesis rate in the *sample* of plants growing at 20°C was greater than that for the *sample* of plants growing at 10°C, but it also tells us *how confident* we can be that there was really a difference between the populations. In this case the statement is informing us that based on the data in the samples there is only a 1.6% chance (0.016 probability) of getting

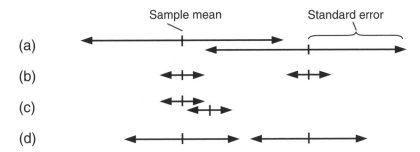

Figure 2.4 Comparing samples from two different populations. The arrows indicate the range of values where the population mean is most likely to lie. (a) Large difference between sample means but large standard errors; the population means could easily be the same. (b) Large difference between sample means and only small standard errors; the population means almost certainly do differ. (c) Although there are only small standard errors, there is only a small difference between the sample means; the population means could therefore easily be the same. (d) Fairly large difference between sample means but fairly large standard errors; it is difficult to be confident from looking at the diagram that the population means really differ

two such dissimilar samples if temperature really had no effect. This would be good enough for us to conclude that there probably *is* a difference between photosynthesis rates of seedlings growing at 10°C and seedlings growing at 20°C in general (not just in our samples). Notice, however, that the result is only that we can say this is *probably* true, not that it is *certainly* true. This is always the case with statistical tests.

Null hypothesis and *P*-values

If you carry out a statistical test by computer it is usually fairly simple to obtain the answer, but you should also be clear about what question this is the answer to. Statistical tests start with a statement called a *null hypothesis*, which is always along the lines 'there is *no* difference between the populations', or 'there is *no* relationship between the measurements'. The null hypothesis is often given the symbol H_0. The question the test is really addressing is, How likely is it that the null hypothesis is true? And the answer given by the test is usually in the form of a *P-value* (or probability).

The *P*-value given by the test tells us the probability of getting such a result if the null hypothesis were true. A high *P*-value therefore tells us that the null hypothesis could easily be true, so we should not conclude there is a difference. A low *P*-value tells that the null hypothesis is unlikely to be true and we should therefore conclude that there probably is a difference. Hopefully you can follow this if you read this slowly but it is quite cumbersome to recreate this line of thinking every time you want to interpret what a *P*-value is telling you. For all practical purposes it is easier to remember a simple rule:

- Low *P*-values indicate we can be confident there is a difference.

- High *P*-values indicate we cannot be confident there is a difference.

All statistical tests work this way round.

Alternative hypothesis, one-tailed and two-tailed tests

As stated above, the null hypothesis – a statement of what the test is actually testing – is always something like 'there is *no* difference between the groups we are comparing'. The exact format of the null hypothesis will depend on the type of test we are using. We can also make a statement about what else might be the case, which usually takes a form like 'there *is* a difference'. This is called the *alternative hypothesis* and is often given the symbol H_1 or H_A. Between them, the null hypothesis and alternative hypothesis should cover all eventualities.

A typical experiment might be trying to discover whether adding fertilizer will improve the yield a farmer gets from his crops. We might be quite confident that it will, but we must also allow for the possibility that it *could* reduce yield, otherwise we will bias the result of the test in favour of finding what we expect or hope to find. We therefore need to use a test that allows for both possibilities, a so-called *two-sided* or *two-tailed* test (i.e. a test with a two-sided alternative hypothesis).

Our alternative hypothesis would therefore be 'fertilizer produces a *difference* in yield', which allows for the possibilities that fertilizer *either* improves *or* reduces yield. Together with the null hypothesis 'fertilizer does not affect yield', all possibilities are therefore covered. If the test tells us that we can conclude there is a difference, *and* the sample means show that yield was greater in the fertilized treatment, *then* we can conclude that fertilizer improved yield.

In some situations the only possibilities are (i) that there is no difference (the null hypothesis) and (ii) that A is *greater* than B (the alternative hypothesis). An example might be when we know that at a particular site, without fertilizer, the median yield per plant is zero and we want to know whether adding fertilizer improves this. (The *median*, like the *mean*, is a way of describing a 'typical' plant, but is preferred in some situations; Section 13.1). Here, it is not *possible* for adding fertilizer to reduce the median yield. In such cases we should use a *one-sided* or *one-tailed* test, (i.e. a test with a one-sided alternative hypothesis). Our alternative hypothesis would therefore be 'fertilizer *increases* median yield'. Together with the null hypothesis 'fertilizer does not affect median yield', all possibilities are then covered.

It is uncommon to need to use one-sided tests in practical environmental or biological science because it is usually *possible* for the groups we are studying to differ in either direction, though previous experience may tell us which is most likely. The tests described in this book are therefore two-tailed tests. However, it is simple to turn *t*-tests, *F*-tests, correlation and non-parametric equivalents of them into one-tailed tests should you ever need to.

The *null hypothesis* for the test will not be changed but the *alternative hypothesis* stated for the test should be replaced by a one-sided alternative hypothesis, e.g. H_1: mean of A > mean of B, or H_1: A is *positively* correlated with B. The calculations are not affected at all but the *P*-value produced by the test should be divided by 2. These are the only changes required.

How confident do we need to be?

Scientists are usually keen to find a particular result to support their pet theories, whether it is that populations do differ, or that they don't. Recall that experiments almost always end up stating (i) what seems to be the case *if we assume* our samples are like the rest of the populations they came from, and (ii) *how confident* we can be that our samples were like the rest of the populations they came from. To avoid arguments about what constitutes confident enough, a convention has built up:

P-value	Conclusion
>0.05 (e.g. 0.12)	Insufficient evidence to say that the populations differ, or *no significant difference*
≤0.05 (e.g. 0.006)	Enough evidence to say the populations do differ, or *a significant difference*

The term 'significant' should only be used in scientific writing to refer to results of statistical tests with a *P*-value ≤0.05. Terms like 'highly significant' or 'very highly significant' are sometimes used to describe results where the *P*-value is < 0.01 or < 0.001 respectively, but there is no strict convention on this. The smaller the *P*-value, the more confident we can be that the populations do differ, but to express just how confident we can be it is best to quote the actual *P*-value along with your conclusion. To avoid confusion it is better never to use the word 'significant' in the normal sense of 'important' in scientific writing.

Is it sensible that the scientific community accepts things as probably true when there is still up to a 5% (= 0.05) chance that they are wrong? Certainly if I was going to invest £1 million in a project, I would be looking for better evidence than this. In any particular situation we can decide for ourselves what degree of uncertainty we are prepared to accept in a conclusion. You do not *have* to accept the conventional 5% cut-off point (i.e. *P* = 0.05) as good enough.

However, bear in mind that if science routinely insisted on stronger evidence, the cost of research would soar.

Would this be justified? I think the best answer I have heard to this was to ask the question, Do the environmental and biological sciences have a problem with the reliability of scientific research because of this? Not really, as far as I can see. Most important findings are eventually checked out by other workers, so their reliability is tested at that stage anyway. My own feeling is that you should be cautious about accepting any particular result (by definition none are 100% reliable), but that the system as a whole works well as it is.

2.6   Limitations of statistical tests

Even if we conduct an experiment perfectly, collect the data, and analyse it with the correct statistical test, we will be left with uncertainty in our result through no fault of our own. This is because we are trying to say something about large populations when really all we know about are the few individuals in our samples.

Most of the time statistical tests will lead us to the correct conclusions about the populations we are studying. However, there is an element of chance. We might have measured individuals in our samples that were unusually small or large. In this case, even though we have followed the recommended procedures, we could still reach the wrong conclusion. Wrong conclusions are called *type I* and *type II errors*. Making one of these errors does not mean that *you* have done anything wrong, but they are important concepts to know about to understand what statistical tests are really showing.

Although there will *always* be some risk of making type I or type II errors when we use the results of statistical tests, like most things with the word 'error' in them we should do what we can to minimize them. The principal way we do this is by making a lot of measurements or observations (i.e. gathering a lot of evidence). I will cover selecting a suitable sample size in Section 4.2.

Type I errors

If we conclude from a statistical test that there was a difference between the populations we were studying when really there was not, we have made a *type I error*.

Supposing a statistical test gives us a *P*-value of 0.034 (a significant result). We have to base our conclusions on the evidence of our samples, so we would conclude that there probably was a difference between the populations. However, we could have obtained such a result either (i) because the populations really did differ, or (ii) because although the populations did not differ, by chance the individuals in our samples were not typical.

If (ii) happens to be the case we will have actually reached the *wrong* conclusion; we will have made a type I error. In a practical study we will never know if we have reached the wrong conclusion in this way, but we can say what the chances of this are. In this case the chance we have made a type I error in reaching our conclusion is 3.4% (because $P = 0.034$). A 'significant' result only means we have good *evidence* of a difference, not proof.

Type II errors

If we do not conclude from a statistical test that there was a difference between the populations we were studying when really there was, we have made a *type II error*.

Supposing a statistical test gives us a *P*-value of 0.091 (a non-significant result). We would *not* conclude that there is a difference between the means of the populations. However, we could have ended up with a result like this either (i) because the populations really did not differ, or (ii) because although the populations did differ, by chance the individuals in our samples were not typical, or (iii) because although the populations did differ our experiment was not very thorough (i.e. we didn't take enough measurements to make a good assessment).

If (ii) or (iii) happens to be the case we will have again reached the wrong conclusion; we will have made a type II error. In practice it is difficult to quantify the chances that we have made a type II error and so usually this is not done. However, you should at least appreciate that non-significant results can be in error in this way.

A 'non-significant' result means that we have *little or no evidence* of a difference, not that we have evidence of no difference.

Power

For completeness here I mention another term, the *power* of a test. This is closely linked to type II error; if you have a more powerful test, you are less likely to make a type II error. Like the chances of making a type II error, it is not usually quantified but it is important to know about it as a concept. It is best to think of the power of a test as the thoroughness of a study.

The more powerful a test, the more likely you are to get a significant result if the populations really do differ. The power of a test can be increased by increasing the sample size and, as the name suggests, it is usually desirable to have as powerful a test as you can.

The actual terms 'type I error', 'type II error' and 'power' are seldom used outside statistics textbooks; however, it is important to appreciate at all times that both positive and negative results of scientific research generally contain uncertainty.

Non-significant results

A final thought before we leave this. You will sometimes see scientists quote negative results as part of their argument, as in this example:

> Lead concentration affected plant growth ($P = 0.02$) but iron concentration did not ($P = 0.08$).

To *conclude* that the different iron concentrations in the study did not affect plant growth is wrong here. On the whole, the evidence is pointing towards the fact that they did, but it isn't strong enough to conclude this (because $P > 0.05$). Now that you have read about type II errors you should recognize that this non-significant result might be obtained either because there really was no effect of the range of iron concentrations used, or because by chance the samples happened to be atypical of the rest of the populations, or because the scientists did not carry out a thorough enough investigation to observe the effect. Very well done if you really did spot this.

For this reason, statisticians usually advise against using non-significant results to conclude there was no difference. Certainly you should try not to do this, by designing your experiments carefully or carrying out a further experiment (e.g. test iron concentrations again using a larger sample size). But in practice this is not always possible; perhaps you don't have facilities for handling bigger samples, or the question does not arise until your opportunities to do practical work are already finished. In this case there is, at least, a measure of the sensitivity of your test that you could quote along with your observations. This is called the *least significant difference*, and we will look at this in Section 7.3. In effect, it allows you to make a statement such as:

> There was no evidence that iron concentration affected plant growth but if it did it was probably by no more than $x\%$.

3

Before You Start

3.1 Introduction

I tried to emphasize in Chapter 1 that analysing a well-planned experiment or survey is easy. On the other hand, getting the information you want from a badly designed experiment or survey will probably be difficult, and might turn out to be impossible. People who struggle with the statistics part of a project are almost always people who did not bother to plan well. Numeracy skills do not feature very much when we are simply *using* statistics these days. After all, a computer is probably going to do the calculations for you.

The key to a happy life is to deliberately try to frame your research questions in a format suitable for one of the statistical tests you know about. It is much more difficult, and sometimes impossible, to take a set of experimental results and *then* try to identify which technique is needed to analyse them.

Supposing you read in a statistics textbook that you can compare the mean yields of wheat plants growing in several different fertilizer treatments using a test called one-way ANOVA. The examples given will describe using sets of experimental field plots that differ only in the type of fertilizer treatment, and samples of plants will be selected and yields measured in the same way for each treatment. The numbers are then analysed using this test, a result is produced and you are told how to interpret this in terms of the effect of the fertilizer.

Now, perhaps you are interested in whether different pH treatments affect the time it takes for bacterial colonies to double in size. Can you set up an experiment comparable to the wheat/fertilizer trial in the textbook? Well, in place of field plots you can have sets of Petri dishes. In place of fertilizer treatments you can have growth media that differ only in pH. And instead of measuring yields you can measure the time it takes colonies to double in size, in the same way for each pH treatment. Although you are studying a very different subject to the example in the textbook, you could set up an experiment with a very similar layout. Then, all you have to do is enter your own experimental data into a computer and you can follow the example in the textbook for guidance about interpreting the result. If you go about things in this way, statistical analysis of experiments is really very straightforward.

Below I describe the kinds of statistical methods available and some general points to consider when planning a study. A list of worked examples in this book is given in Appendix E. Once you have a clear idea of which statistical method you will use and how you are actually going to collect measurements or observations, further details of the formal procedures for setting out an experiment or survey are given in Chapter 4.

3.2   What statistical methods are available?

A wide range of tests have been developed for different purposes. I only cover a selection in this book but they are chosen to cover most of the types of research question you might want to answer in environmental or biological science. However, this might require you to break down complex questions into a set of simple questions first. This is where planning an experiment around statistical tests becomes vital to the success of your work.

The methods in this book can be put into five categories, described below. Not all are tests as such; some just help to identify patterns in data by presenting them in a particular graphical way. When designing an experiment or survey, consider whether your research question is, or could be, organized in an appropriate way for the various techniques.

Statistics for describing populations

The statistic used most commonly to describe a 'typical' member of a population is the *mean*. The statistics used most commonly to describe the spread of values within the population are the *standard deviation* and *variance*. An alternative to giving the standard deviation is to give the *coefficient of variation*. The mean that we calculate comes from a sample, so when we use this as an estimate of the mean of a population there is a 'margin of error' in it. It is usual to also give this 'margin of error' for the mean by quoting the *standard error* or 95% *confidence interval*. These statistics are all described in Chapter 2.

If it is not possible to calculate the mean because some values in the sample are off-scale, or because the values are actually ordered categories rather than measurements, it is sometimes possible to give a 'typical' value by using the *median* instead of the mean (Section 13.1). The spread of values may then be indicated by the *interquartile range* (the range between the minimum and maximum values remaining after removing the lowest 25% of values and the highest 25% of values from a dataset).

Tests that compare 'typical' members of populations

To compare the means of two populations we could use a *t-test* (Chapter 7), e.g.

Are the mean leaf lengths the same for a species of tree growing in two different countries?

To compare two proportions we could use a *chi-square test* (Chapter 12), e.g.

Are the proportions of germinated seeds the same in two different watering treatments?

To compare the means of several populations we could use *one-way ANOVA* (Chapter 8), e.g.

Are there any differences between the mean yields of maize per hectare in different countries in East Africa?

To study the combined effects of two factors we could use *multiway ANOVA*, e.g.

What are the combined effects of numbers of cars, and temperature on atmospheric SO_2 concentrations?

ANOVA can be extended to study the combined effects of three or more factors; suppose we also wanted to know whether time of year affected SO_2 concentrations. While it is theoretically possible to study any number of factors at once with this technique, it becomes very difficult to interpret the results satisfactorily when more than three factors are involved (Chapter 8).

Analysis of repeated measures

To compare the same groups of individuals at a series of points in time, we need *analysis of repeated measures* (e.g. comparing distances walked by different species of deer by observing a number of individual animals daily for 6 weeks).

People are often disappointed by what they can learn from repeated measures. It may be, at the end of their experiment, all they can say is that the groups were not the same, and a single measurement made in the middle of this time might have been enough to tell them that. It is therefore important to read about analysis of repeated measures before deciding to collect this kind of data (Chapter 11).

Non-parametric tests

Sometimes the mean is not a good representation of a 'typical' member of the population because a few values are unusually large or because the individual data points are not actually values but categories (e.g. very small, small, medium, large, very large). It might still be meaningful to compare 'typical' members of two or more populations, but we could not calculate the means because there are no actual values. In these cases we use the *median* to represent a typical value and analyse the data using *non-parametric tests* (Chapter 13).

Tests that compare distributions

We saw in Section 2.2 that the way individual values were spread within a population was called a *distribution*. If we have mostly high values and a few low values this is one type of distribution; if we have mostly low values and a few high values this is another type of distribution.

To compare how widely spread the values are in two populations which both have Normal distributions (Section 2.2) we could use an *F-test*, e.g.

> Are the distances travelled by lorries within the Netherlands more or less variable than the distances travelled within the UK?

In technical terminology an *F*-test compares the variances of populations (Chapter 7).

To test whether two populations have the same shape of distribution we could use the *Kolmogorov–Smirnov test*, e.g.

> Are the distributions of stomatal apertures on bean leaves from two different atmospheric pollution treatments the same shape?

This is a type of non-parametric test (Chapter 13).

If we have sets of counts of how many individuals fall into each of a number of categories, we may want to examine whether individuals are divided between categories in the same proportions within different populations. For this kind of comparison of distributions we use a *chi-square test*, e.g.

> Do we get the same proportions of blue, white and pink flowers for milkwort plants growing on peaty, clayey and sandy soils?

Chi-square tests come in several different forms. They can be used to test whether distributions are the same, or whether a distribution conforms to some theoretical pattern. Chi-square tests can also be used to compare proportions of individuals with a particular characteristic, i.e. distributions where each individual falls into one of only two possible categories (Chapter 12).

Tests for relationships between measurements

Tests for relationships between measurements are used when we are measuring two or more different characteristics of the same group of subjects and we want to know if they are related, e.g.

Are height and weight related?

To test whether there is a linear relationship between one type of measurement and another, made on the same individuals, we use *correlation*, e.g.

Is there a linear relationship between head diameter and IQ?

By 'linear relationship' I mean that an increase of one unit in one of the measurements is always related to the same size of increase or decrease in the other, e.g. if each additional 1 kg of body weight in adults is associated with an additional 80 kJ in dietary intake, this is a linear relationship (Chapter 9).

If we have established that two measurements are linearly related, we can derive an equation to predict one from the other using *linear regression*, e.g. in the example above we might get an equation such as

$$\text{daily energy intake} = 5000 + 80 \times \text{weight}$$

With linear regression we can also compare whether the same relationship applies in two different populations, e.g.

Do men and women have the same relationship between body mass and daily dietary intake?

Or examine how one measurement relates to several others (Chapter 9), e.g.

How is river flow rate affected by size of catchment, rainfall, and percentage forest cover in a catchment?

If we believe there is a relationship between two measurements but it is not linear, we can still test for a relationship using *Spearman's rank correlation* (Chapter 13). We can also often produce equations relating one measurement to another, when the relationship is not linear, using *polynomial regression* or *nonlinear regression* (Chapter 9).

Multivariate methods

Multivariate methods make use of *several* pieces of information about each individual in a sample at the same time, to study similarities and differences between them (e.g. for grass plants we might study their heights, shoot weights, root weights, root lengths, numbers of leaves, and shades of green).

To test for differences between two or more populations, based on several characteristics (rather than just one characteristic such as height) we could use *multivariate ANOVA* (Chapter 10), e.g.

> Do four river estuaries have the same invertebrate populations, considering numbers per litre of 10 different invertebrate species?

To identify which characteristics or which combinations of characteristics vary most between individuals, we could use *principal component analysis* (PCA), e.g.

> Which are the best of 20 suggested indicators to study, to give an early warning of pollution in fresh water lakes?

A typical outcome might be that we find that samples from lakes known to have different degrees of pollution showed a wide variation in dissolved oxygen and nitrate contents, but measuring the other suggested characteristics would tell us little or nothing extra.

Identifying the most useful characteristics in this respect can help to focus future research. The results of the first stage of the analysis are sometimes plotted graphically and occasionally distinct groupings of individuals appear in the graph, which might not previously have been thought of. For instance, it might turn out that some of the plants in a study had very large above-ground parts in relation to their roots, and careful thought may reveal that these tended to be from wetter sites. In this way it might be discovered that this species' morphology was strongly influenced by soil water content. Although the results of PCA can sometimes reveal interesting new facts, they are often not clear-cut and there is no yes/no answer produced by the analysis (Chapter 14).

To determine whether some individuals are more alike than others, considering several characteristics together, we could use *cluster analysis*. Suppose we have grass plants from several mountains, several coastal sites and several agricultural field sites. Considering several characteristics (length, height, weight, etc.), we may ask:

> Are the plants growing near the coast more like each other than they are like those
> on the other sites, and are the plants growing in the fields more like the mountain
> plants or the coastal plants?

Cluster analysis produces a graphical result where similarities between individuals and groups are shown on a scale from totally the same to totally different. It does not produce a yes/no answer (Chapter 15).

3.3 Surveys and experiments

Environmental and biological studies can be broadly classified into surveys, in which we are aiming merely to observe and possibly compare populations, or experiments, in which we aim to control some aspect of the populations or their environments to find out how they respond. In both cases we will actually be studying samples made up of relatively few individuals, which we hope are representative of the populations as a whole. Chapter 4 gives further information about selecting representative samples and how to choose a sample size.

Surveys

Surveys in environmental and biological sciences are studies of naturally existing populations (e.g. the rubisco content of the plants growing in certain areas, the hardnesses of the rocks on certain mountains, the PCB concentrations in certain streams, or the numbers of species of bats in certain caves). Textbooks on surveys sometimes refer specifically to studies on people. Although the statistical analysis is very similar in both cases, these books often focus on issues such as whether people are likely to answer certain questions honestly or whether some social classes of people are more likely to be at home if we phone at a certain time of day. Clearly these are not relevant in most environmental or biological studies. In all cases, though, we aim to select a sample that is truly representative of the populations we want to study.

We therefore have two key concerns in survey design:

- We want to give every individual in the population an equal chance of being in our sample.

- When we make our measurements or observations we want the individuals to be behaving exactly as they would if we were not there.

Surveys may be carried out simply to collect data on some topic (e.g. fish stocks in different areas of the North Sea), or they may be carried out to compare different naturally occurring populations using statistical tests.

Experiments

Experiments are studies which involve either (i) comparing how individuals of the same type behave in different environmental conditions, or (ii) comparing how different types of individuals behave in some specified environmental conditions. As with surveys, they involve using samples to study populations. In experiments we will be using *samples* of individuals of a specified type in specified environmental conditions to infer how individuals of these types in general, the *populations*, would behave in these conditions (e.g. how the rubisco content of plants would be affected by atmospheric CO_2 concentration, how the breaking strength of rocks would be affected by their shape, or how a certain species of invertebrate would respond to different concentrations of PCBs). We usually refer to the different types of individuals we are comparing, or the different environmental conditions we are comparing, as *treatments*.

Key concerns in experimental design are:

- If we are studying the effects of *different environments*, there should be no differences between the samples of individuals used in the different experimental treatments at the start of the experiment. If we are comparing *different types of individuals*, they should all be placed in the same environmental conditions during the experiment. This way it will be clear exactly what has led to any differences that are observed during or after the experiment.

- The experimental treatments must be identical in all respects other than those we are interested in studying. For example, if we are studying the effects of temperature on some organism, then light, humidity, pH, etc., must be the same in all of the treatments, so that we can attribute any effects solely to temperature. If we are comparing different types of individuals, they must be the same in all respects other than those we are interested in studying. For example, if we are comparing the milk yields of two breeds of cattle, the sample animals from each breed should be the same age, fed on the same diet, milked at the same time of day, etc., so that we can attribute any differences solely to breed.

Survey or experiment?

Both surveys and experiments have advantages and disadvantages. Surveys often give us good evidence about whether populations really do differ or whether certain relationships really exist in the natural environment. However, the environments of the populations we are comparing may differ in many respects (e.g. light, latitude, pollutants, altitude, distance from the sea), so often we cannot be sure from a survey *what* has caused the differences or relationships that we have observed.

Experiments are usually intended to test theories produced in the light of other experiments or in the light of a survey. We can usually be confident at the end of an experiment what has caused any effects or differences we have observed, because we have controlled all the other possible influences in the way we have carried out the study. The downside to experiments is that we have usually created an artificial environment or an artificial population, so we cannot be sure that what we have observed in the experiment also happens in the natural environment. Combining observation of the natural environment through surveys, and then testing theories about how it works through experiments, is a powerful approach to research.

The statistical analyses of surveys and experiments are generally the same. For instance, a *survey* might involve measuring the lead contents of faeces from flocks of sheep grazing in different parts of the country, or an *experiment* might involve measuring the lead contents of faeces from groups of sheep fed different, carefully controlled diets. In both cases we could use one-way ANOVA to compare the mean results for the different groups.

3.4   Designing experiments and surveys – preliminaries

In this section I cover some general points to consider in the initial stages of designing an experiment or a survey. These will help you to decide whether you really have the resources to carry out the study you would like to, or whether you would be better to do something simpler but do it well. The actual layout and size of a study are covered in Chapter 4.

Decide what questions you are asking

If you are clear about the question you intend to answer, it will lead you directly to a particular statistical test or method, or at least tell you that there is no suitable test available so you can think again. For example, if you want to know whether the mean beak lengths of two populations of purple sandpipers differ, the logical choice is a *t*-test because *t*-tests are designed to test for differences between the means of two populations. But suppose you have only a vague idea of the question, e.g.

I want to look at beak lengths of purple sandpipers.

Intuition will tell you to go and make some measurements of beak lengths of these birds, but if they are not collected in the right format, you will not necessarily be able to carry out a statistical test on the measurements afterwards.

Now, it may well be that having posed your question clearly, you cannot find any statistical techniques in this book to answer it. This is no bad thing, because you then have the option to modify your question or seek information elsewhere, but you have not wasted any time making the wrong measurements in the field. (I'm not sure how you would measure beak lengths of purple sandpipers but my guess is it would be quite time-consuming.) Once you have managed to match a question that interests you with a statistical technique that can answer it, and that you feel capable of using, you can proceed. This reconciliation of what question you would *like* to answer and what questions you have the *ability* to answer, is a vital part of experimental design.

Design your experiment or survey around a statistical method

Having selected an appropriate statistical method, use the examples in textbooks as blueprints for your own experiments. Try to see the parallels between their situation and yours. You may feel all this is rather expansionist. Statisticians are trying to dictate how environmental or biological science should be done, when statistics should be there just to help you to sort out the answers at the end. However, you will find this view is not very helpful to you. For one thing, statistics isn't somebody else's subject, it's a part of ours. For another, ask yourself these questions: Are you going to have to do statistical analysis on your results whether you like it or not? And would you prefer this to be straightforward? If so, then design your experiment or survey around a statistical method and save yourself a lot of work!

Once you have a bit of experience with designing experiments or surveys, you will often be able to match your design to a statistical technique in your head and so design your data collection in an appropriate format. However, when I come to a situation that requires something I haven't tried before (and when you are starting out I suppose this means any situation), I find it useful to check I will be able to analyse my results by making up dummy results and trying to analyse those first. What I am testing here is simply that it is possible to analyse a set of results like the ones I plan to collect, using the computer package I have available and that I am going to get a useful result.

There are several reasons why you might not be able to carry out the analysis as you have planned.

The statistical package you have available is limited in the kinds of analyses it can carry out
Most statistical packages will cope with most of the methods described in this

book (and a great deal besides). In my experience though, most packages turn out to have surprising omissions – methods that are used commonly in my own research area but presumably are not widely used elsewhere. If you plan to do your statistical analysis on a computer, check that you have access to a program that will handle your proposed analysis.

The computer carries out an analysis but the results are presented in a form you can't understand
Different programs produce the results in different layouts, though they are usually recognizable. Some may give several different versions of the result where this book describes only one. A useful tip here is to try out an example from this or another textbook and compare the results with those given by the computer. Still, if you really can't understand the output, and no help is available, there is little point carrying out an experiment that will generate this kind of result. It would be better to redesign your experiment around a simpler statistical test.

It is not possible to complete the calculations for the test
In this case the computer will return an error message instead of a result. This can occur because the sample size is too small (Chapter 4). However, it can also occur because of a lack of *balance* in the experimental design. When we are studying the combined effects of two or more factors (e.g. light and tempera-ture), we need to include treatments in our experiment that are combinations of the different factors (e.g. high light + low temperature). If we haven't included the right combinations, we can run into problems with the statistical analysis. In some complex designs this can be difficult to predict, but it would be revealed by trying the analysis using dummy data. Usually, however, we can avoid this by using a *factorial* set of treatment combinations (Section 8.4).

Consider what you already know about the populations you are studying

Hopefully you will do some reading before starting any experiment or survey to make sure you are not simply duplicating work that has already been done. Scientific papers usually give not only their findings, but also information about what sample sizes were used, experimental conditions, how many measure-ments were made in a day, etc. Though your own study will probably be unique in some respect, it will probably have a lot in common with other studies in other respects. For example, you may be studying effects of pH on growth of some fungus, whereas someone else has studied the effects of nutrient concen-tration on the same fungus.

You will learn a lot by noting how they conducted the experiment. They might mention the need to ensure that all of the treatments were applied to fungi from the same batch, or that the same number of cultures of each treatment were placed into each incubator. Although you are studying a different effect, you can make use of advice like this in your own work. Look also at the 'margin of error' they got in their results (usually shown as the standard error on graphs or in tables). If you use the same number of *replicates* (i.e. the same number of individual measurements in each sample) and you are conducting a similar experiment, you will probably get a similar 'margin of error' in your results. If you try to cut corners by using fewer replicates, you will very likely end up with a larger 'margin of error'. Would this still make your experiment worthwhile?

Although you may skim over the details of methods and ignore standard errors when you are reading scientific literature to find out what the authors discovered, when it comes to designing your own experiments or surveys, these are a fantastic source of ideas and information. Remember also that some of the people around you probably have useful experience too, particularly if you are working in a university or research institution. Ask your co-workers about their experimental designs. They have the advantage over published work that they can also tell you about things that went wrong and why.

Be sure the design makes practical sense

This has nothing to do with calculations or formulae but is often the aspect of an experimental or survey design that requires greatest thought. Even though you might successfully make a set of measurements and carry out a statistical analysis, the answer you get is not necessarily telling you what you want to know if you have not thought about the physics and physiology of your study. Here are some examples.

Example 1: Carrot crop

Suppose I want to study how a carrot crop develops over a 6 week period; just in case you didn't know, carrots develop underground, so you have to dig them up to see them. I decide to use five small experimental plots with 36 plants in each and make sure these plots are randomly located within the study field. Each week I dig up six randomly selected plants in each plot and make measurements on them. This has most of the elements of a good design but fails to take account of how my presence is going to affect the plants I am measuring.

Firstly, I might damage neighbouring plants each time I dig some up; secondly, I will be disturbing the soil, which might affect its water content or allow weed seeds to germinate; and thirdly, each week the remaining plants in my survey plots will be a little more spread out, which will probably improve their growth. The plants in the rest of the field (i.e. the population I hope to represent), however, will stay at their original planting density throughout. There would be little point going ahead with this survey because it would not give me an accurate picture of how the crop usually grows.

Example 2: Nitrogen deposition

If it had been observed that nitrogen deposition varied greatly between the top and bottom of a hill, we might want to test whether this affected plants' growth. We might try growing some plants near the top of the hill and some near the bottom and comparing their growth. We would generate a set of results which we could compare using a *t*-test. However, this would not necessarily be telling us the effect of differences in nitrogen deposition, because any differences we found might also be due to differences in temperature, wind speed or some other factor.

When several different factors could have been responsible for the effects we observe, we say that these factors are *confounded*. Generally speaking, having confounding factors in an experiment or survey is a bad thing because we will not know the real cause of any effects we observe. Experiments must therefore be designed thoughtfully to avoid them. If we sampled individual carrots from 30 different, randomly selected locations in the field on each occasion, rather than from 5 small plots, our previous sampling would not affect the plants we dig up each time.

Situations such as the study of nitrogen deposition present a common problem to which there is often no ideal solution; we want to know the effect of only one factor but we don't want to affect the other environmental conditions while we study it. Possibilities include trying to simulate the conditions in controlled environment cabinets, or growing plants at both the top and bottom of the hill, each with both high and low nitrogen deposition treatments, which we control in some way to simulate the amounts found naturally at the top and bottom of the hill. By including all four possible combinations of top or bottom of the hill + high or low nitrogen deposition, when we come to analyse the data, we will be able to identify the effects of nitrogen inputs, as opposed to other effects of position on the hill. This is an example of a factorial design (Section 8.4).

Factors which are confounded in an experimental design cannot be separated at the stage when we analyse the data, so we should try to avoid problems like these by thinking carefully about the experiment before we start, and making sure that we do not allow other factors to be confounded with the factors we are trying to study.

Consider what you will do about missing values

In many experiments or surveys the number of measurements or observations we eventually get turns out to be slightly less than we intended because things don't go according to plan. The reason these missing values have occurred influences how we should handle the rest of the data. Never simply enter zeros in place of missing values when you know that the real values were very unlikely to be zero. These values will be treated by the statistical tests as if you had actually measured values of zero and so give you a very misleading answer.

Missing values which are 'typical' values can occur because

(a) You drop something

(b) You lose a piece of paper

(c) Somebody has 'borrowed' one of your beakers

(d) When you arrive at the site you only have 17 plastic bags

Values that are missing for reasons like these generally do not present a problem. Most statistical packages allow us to enter this kind of missing value as an asterisk or other symbol. Essentially they are assumed to be like the average of the other values in the same group. However, they can present a problem if you are using a basic statistical package that does not permit you to enter missing values, or you have already cut the number of measurements down to the absolute limit required to do the calculations required for the test (Section 4.2).

You can check for either of these possibilities before collecting your data by using dummy data, as suggested above, and then deleting a few values. If it turns out that missing values are going to prevent you from analysing the rest of the data, try increasing the number of measurements you make, checking whether you have access to a better statistical package, or at the very least, being very careful!

Missing values which are not 'typical' values can occur because

(e) It started to rain heavily so you were only able to trap the nocturnal rodents until 1:00am instead of all night as intended.

(f) The small plants died so you could not measure their transpiration rates.

(g) The highest values of soil hardness were off the scale of your instrument.

Values which are missing for reasons like these are more difficult because we cannot assume they are like the rest of the values in that group. They should not, therefore, simply be entered using the 'missing values' symbol available on statistical packages. Assuming you don't have the option to start again, you have to do something, so here are some possibilities.

In case (a) you could redefine the population your study relates to (i.e. it is now: a study of 'nocturnal rodent behaviour up to 1:00 am', rather than 'nocturnal rodent behaviour in general') and analyse the data you have.

In case (b) you could report that a certain number of plants died, and also redefine the population you studied to be transpiration rates of *surviving* plants. You could then go ahead and analyse the data from just the plants that did not die. You might also try to analyse the counts of numbers of plants that did die to test for differences between the different treatments using a chi-square test (Chapter 12), and thereby gain an additional result from your experiment.

In case (c) it is pointless to redefine the population. It would end up as something like 'the average hardness of all soil samples with a hardness less than 3 MPa', which is meaningless. The main problem here is that if we do not know *all* of the values, we cannot calculate the mean. Many of the tests described in this book compare sample means, so we would not be able to use them. Most of the *non-parametric* tests could still be used in this case (Chapter 13), but they are quite limited. Off-scale values in a dataset can therefore present considerable problems, so it is worth planning ahead and trying to avoid them in the way you carry out your measurements.

Avoid pseudoreplication

Statistical methods are based on the assumption that we have measured or observed samples that are representative of particular populations. The samples must be representative in terms of both 'typical' values in the populations and the *distributions* of values in the populations. Otherwise the results given by the tests will not be accurate.

Consider the situation in Figure 3.1. The variation between individual measurements from scheme (b) will always be less than from scheme (a). If we compared samples from two different fields using sampling scheme (b) we would be much more likely to get a significant result in a statistical test than if we used scheme (a). This does not mean that scheme (b) is a more sensitive way to do the analysis. It is simply an *incorrect* way to do the analysis, so we would obtain an incorrect result (which might happen to look more favourable from our point of view). Statistical methods *need* information about variability between individuals in the populations we are studying.

We use *replication*, i.e. we measure more than one individual from each population, so that we know something about the variability *inherent* in the population. In scheme (b) we get no information about the variability within the field. There will be some variation in the results but this is just variation due to

(a)   (b)

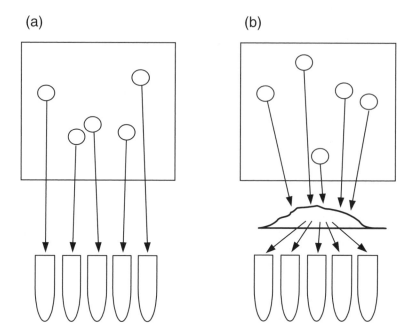

Figure 3.1   Two possible sampling schemes for characterizing the soil in a field. (a) The correct procedure: samples of soil are taken from random locations in a field and analysed separately. (b) An incorrect procedure: samples are taken from the field and thoroughly mixed; samples are then taken from this mixture and analysed separately

the repeatability of the measuring process employed. Sampling scheme (b) is called *pseudoreplication* because it does not provide real replicate measurements required to use statistical tests.

Ensure that individual measurements are independent

A further requirement of statistical tests (unless otherwise stated) is that the individual measurements within the samples are *independent* of one another. In other words, it is assumed that each individual we measure would have behaved the same, whichever other individuals had been in the sample.

Consider an experiment to measure weight gain of starlings fed on two different diets; two cages are used with five birds in each. The birds in one cage are fed on a normal diet, those in the other cage have a diet supplemented with a new formulation, Tweeto. On the face of it, we have five replicates (five birds in each treatment that we can measure individually at the end). However, this design is flawed because some of the birds in each cage may be dominant and leave little food for the others. The weight gains of the individual birds are therefore not independent of the other members of the sample.

In another experiment we might want to compare the effects of soil type on growth of some plants. Suppose we set up one plant pot with a sandy loam soil and one with a clay soil, and place 10 plants in each. Unfortunately, there is only so much nutrient, so much water, and so much light available in each pot. Therefore if some plants grow very well, they will affect the growth of the others. The growths of the individual plants in this experiment are also not independent of each other.

The calculations in most statistical tests are based on an assumption that the individual measurements *are* independent. Of course, if you have a big enough pot containing 10 plants, they will not affect each other in any way. But how big does the pot need to be, and how can you ever prove that the individuals you studied were able to function independently? In practice you just have to be reasonable about this. Everyone works within practical limitations but your design does need to be such that one would reasonably expect the members of your sample to have a negligible effect on each other.

A further common cause of non-independence of measurements is when the same set of individuals is measured on several occasions, e.g. measuring the concentrations of a pollutant at the *same* sites every week. Such measurements are called *repeated measures*. Unlike the examples above, this is not necessarily a bad thing, and we might ultimately choose to do this. However, it is a bad thing to do without appreciating the consequences.

The amount of extra information you get from your experiment or survey by making repeated measurements often seems to be disproportionately small compared to the effort you put into making them (Chapter 11).

Once you have got your basic design, always consider whether it can be done a better way before starting work

It is better to throw away a day of ill-conceived planning than several weeks of practical results. If you are not happy with your plans, consider whether you could do the whole thing a different way. Could you ask a simpler question, or test for a relationship instead of a difference? It often also helps to get someone else to look at your design.

3.5 Summary

At the outset of designing a survey or experiment, you will have at least a vague idea about what you want to study. Consider which of the various statistical methods covered in Section 3.2 you could use. If none seems appropriate, consider simplifying your study so that you *can* use one of them. Otherwise you will need to consult other texts or people for advice, but this should still be done at this stage, before making any measurements. Whether you settle for one of the methods described in this book or use another source, read about the test before you start, to make sure it is suitable.

Once you have decided on a statistical method, consider the other points in Section 3.4. The main thing is to make sure the samples of measurements or observations you collect will be genuinely representative of the populations you are intending to study. If this cannot be done, there is little point in proceeding with the study. Once you have decided on a statistical method and worked out how you will physically make the measurements, you can decide when, where, and how many measurements or observations to make by following the guidelines in Chapter 4.

4

Designing an Experiment or Survey

4.1 Introduction

Very few environmental or biological studies are done solely for the interest of the researcher; they are carried out to inform others. Therefore, it is not just important that our results convince us, it is vital that they convince others too, otherwise we have wasted our time. Following accepted statistical procedures for design and analysis of experiments will help us to achieve this. Some of the advice below may seem pedantic and unnecessary, but many of the factors that can bias a result are very subtle and difficult to spot and we may never know that they have influenced our results. Statistical methods have therefore been devised so that we can be sure we have not biased the result, even if we do not know what all the possible sources of bias are. Sticking rigorously to accepted statistical protocols is the best way to make sure our results and conclusions will be readily accepted by others.

Chapter 3 discussed general points to consider at the initial stages of designing an experiment or survey. Once we have decided on a statistical test or technique to use and how to collect measurements so that we will know exactly what is causing any effects we observe, we can start to plan the details of which individuals or points we are going to measure, and when and where we are going to make measurements or observations.

4.2 Sample size

By this stage in our planning we should have decided what kind of measurements or observations we are going to make and which statistical techniques we are going to use. Remember that we will be studying samples of individuals to represent populations, so next we need to decide *how many* measurements or observations to make to get a result that we can be confident is representative of the populations we are trying to study.

We usually use the term *sample size* to refer to the number of measurements made from each population. If we measure the concentration of chloride in 15 bottles of water collected from a lake, each of which contains 100 ml, the sample size is 15, not 100 ml. Alternatively the term *replicates* may be used, particularly when we are comparing several populations. If we measure the chloride concentration in 15 bottles of water collected from each of 5 different lakes (75 bottles in total) we are using 15 replicates.

The effect of changing the sample size you use is to alter the sensitivity or accuracy of a statistical method. Increasing the sample size will improve accuracy or sensitivity. This is really nothing more than common sense. If you are thinking of moving to a new area and you want to know how much houses cost on average, you would not just look at one or two houses in an estate agent's window. You would probably try to look at a few dozen prices, otherwise you might be put off unnecessarily because the first houses you saw were unusually expensive. So it is with experiments or surveys. If we only measure one or two individuals, we might get lucky and reach a correct conclusion. However, if we want to be *confident* of reaching the correct conclusion, we need to be prepared to make quite a lot of measurements or observations.

Disadvantages of having a sample size too small

Result is inconclusive

Our statistical test is likely to give us a rather inconclusive result. If we use very small sample sizes, we will often get 'non-significant' results. In such cases we cannot conclude that the populations differ; it might just be that we have not looked very hard.

Sample is unrepresentative

There is quite a high chance that the random sample we select will not be representative. For example, if we sample the heights of only three people, we might find their mean height is only 1.57 m; whereas if we sample the heights of 1000 people, we are very unlikely to get a sample mean so far from the population mean. Also if we sample the heights of only three people, we might well find that their heights differ by only 3 cm, so our sample would be suggesting that most people are very close to the same height. If we sample the heights of 1000 people, we would almost certainly get a spread of values similar to the spread in the population as a whole.

If either the mean or the spread of values in a sample is very different from the population they came from, we are likely to get a misleading result from a statistical test. The statistical test itself, however, will give no indication that we are being misled by an unrepresentative sample. Therefore, very small sample sizes should be avoided for this reason. When only one or two populations are involved, a minimum sample size of 6 is advisable. For tests comparing more than two populations, this rule of thumb may be relaxed somewhat. For some of the tests, further guidance is given in the section on limitations and assumptions in the relevant chapter.

Calculations cannot be completed

It might not be possible to complete the calculations required for the test. There are absolute minimum numbers of measurements you need to make before the calculations for the different statistical tests can be carried out at all. If you think of each measurement you make scoring one point, technically called a *degree of freedom*, the test will use up a certain number of points (or degrees of freedom) for every question you try to answer. Questions, in this sense, may be: Is there an effect of temperature? Is there an effect of light? And is there a combined effect of temperature and light?

A statistical test may be trying to answer several such questions at the same time, so some tests use up a lot of degrees of freedom. You end up with a number of *residual degrees of freedom*, which has to be greater than zero or the computer will simply return an error message to say the calculation cannot be done. This can be a particular problem in repeated measures and multivariate ANOVA problems (Chapters 10 and 11).

Although it is possible to calculate how many degrees of freedom a test will use, this can be quite complicated. An easier way to check you are making enough measurements is to run your proposed analysis using dummy results. Random numbers generated by the computer will do as a set of dummy results (one number for each measurement you plan to make), but don't use a set of dummy results just consisting of the same value entered the required number of times. If your experimental design doesn't work, the computer will return an error message. If the computer does return a result – no matter what it is – your design passes this part of the test.

Sample size too large

You cannot really have a sample size that is too large. However, you can have a sample size that is larger than necessary. There is little point using a sample size of 1000 if a sample size of 20 is enough to give you strong evidence that two

Box 4.1   Calculating a suitable sample size

Statistical calculations and tests usually require us to input the sample size and produce, among other things, a 'margin of error' such as the standard error or 95% confidence interval (Box 2.2) or the least significant difference (Box 7.1 and Sections 8.3 and 8.4). If instead we start by specifying a 'margin of error' that we are prepared to accept, it is possible to work backwards through the calculations to find the sample size required to obtain this.

For one sample
Box 2.2 explains how to calculate a 95% confidence interval. Expressed as a formula this is

$$95\% \ \text{CI} = (s/\sqrt{n}) \times t_{n-1,0.05}$$

where s is the standard deviation, n is the sample size, and $t_{n-1,0.05}$ is a t-value obtained from tables (Box 2.2), (actually the 95% CI is the sample mean $\pm$ this amount). To use this formula, we need to know the standard deviation and the sample size, and we can use it to obtain a measure of the margin of error we have achieved when estimating the mean of a population, i.e. the 95% CI. If instead we start by specifying what will be an acceptable 95% CI, we can rearrange this formula to give

$$n = (s \times t_{n-1,0.05} / 95\% \ \text{CI})^2$$

This will then give us the sample size, n, that we would expect to achieve our chosen 95% CI.

For two samples
To compare two populations we can use a t-test (Chapter 7). Box 7.1 explains how to calculate the least significant difference for the test – the smallest difference between population means we are likely to detect with a given sample size. Expressed as a formula this is

$$\text{LSD} = t_{2n-2,0.05} \times s \times \sqrt{2/n}$$

where s is the standard deviation of either population (assumed to be the same), and n is the sample size for each sample. If we want to be confident of detecting differences of a particular size, this can be rearranged as

$$n = 2 \times (t_{2n-2,0.05} \times s / \text{LSD})^2$$

where LSD is the smallest difference between population means that we are bothered about detecting. This will give us a value of n, the sample size to use for each sample. The value of n given by this formula should be thought of as an absolute minimum. Because there is an element of chance in statistical tests, we may still fail to detect a difference between two population means if it is close to our chosen LSD. Increasing the sample size above this will push the odds in favour of us detecting differences of this magnitude.

In either of the above cases we need somehow to estimate the standard deviation of the populations and to start by guessing the sample size, n, so that we can look up values for $t_{n-1,0.05}$ or $t_{2n-2,0.05}$. (Box 2.2 explains how to look up t-values.) This gives us a calculated value for n, which we use to get a new value for $t_{n-1,0.05}$ or $t_{2n-2,0.05}$. This in turn gives another new value for n, which we use to get further new values for $t_{n-1,0.05}$ or $t_{2n-2,0.05}$ and so on. In practice it usually only takes two or three iterations to get a value for n that is hardly changing. Note that since n represents the number of measurements to make, we should always round it up to the nearest integer above.

populations differ. The larger sample size will just give you stronger evidence of the same thing.

Choosing a sample size

Unfortunately, there is no way of knowing for sure what the best sample size to use would have been until after the study has been carried out. Most methods for estimating the best sample size require us to make some sort of *estimate* about how variable we expect the populations to be.

In a survey we may want to determine some environmental variable within certain limits (e.g. mean daily discharges of manganese from different factories). Box 4.1 gives a formula to calculate what sample size we would expect to achieve a specified 95% confidence interval. This is essentially a rearrangement of the formula given in Box 2.2 to calculate the 95% confidence interval given a sample of values.

We can use the same logic to derive a formula to estimate the sample size required to detect a specified magnitude of difference between two population means in a two-sample unpaired t-test (Box 4.1). This is essentially a rearrangement of the formula given in Box 7.1 to calculate the least significant difference for a t-test.

The most unsatisfactory aspect of the formulae in Box 4.1, is that you have to *estimate* the standard deviation of the populations you are studying. While you might be able to guess a population mean from experience, most people will have no idea about what the standard deviation is likely to be. In this case you would have to obtain an estimate of the standard deviation by asking around, reading similar examples in the literature, or carrying out a pilot survey using a fairly small sample.

If no information is available and you are embarking on a major study, it probably does make sense to carry out a pilot study. Take a small sample (e.g. 10 individuals) and calculate the standard deviation of this sample. This should be approximately the same as the standard deviation of the population. You can then use this figure in the formulae in Box 4.1 to estimate the sample size you need for your study. Extending this approach to calculating a sample size for trials with more than two populations is also theoretically possible but becomes even more cumbersome to do by hand. However, some statistical packages will do this, which makes it a more viable option.

In practice you are quite likely to be using methods that have been used in a similar way by other workers. A much simpler approach then, is to find somebody else or a published study using very similar populations to yours, and draw on their experience; perhaps they were studying an effect of adding one kind of pollutant and you are studying another. If a sample size of 12 gave them good results then probably it will work for you too. If you want to achieve smaller standard errors in your results, you will need to use larger sample sizes than they did.

4.3   Sampling

Having decided how many individuals to include in our sample, we have to decide *which* individuals these will be. At the heart of all the statistical tests described in this book is the notion that we are interested in very large populations (e.g. puffins in the Hebridean islands, carbon concentrations in stream waters of northern Europe, nitrogen use efficiencies of barley plants growing in some defined conditions, or the microbial population of a field), but our only practical option is to study a small sample of them. We need to be sure that the individuals we select are truly representative of the population we are interested in.

A problem we have as human beings is that very often we have a preferred outcome to the experiment or survey, so if we choose the individuals to observe ourselves, we might bias things in our favour (or at least others might suspect we had). As scientists we should always try to avoid influencing any experiment or survey results ourselves. The way we do this is to select the individuals to include in the sample at *random*. In all of the tests and methods described in this book it is assumed that individual members of a sample are chosen completely at random from the population they represent. Every member of the original population, or every point in an area, or every moment in a period of time, should have an equal chance of being in the sample. It is also assumed that the choice of one member of the sample does not in any way bias the choice of the other members. Such samples are said to be *simple random samples*.

Drawing a random sample

In the case of a survey, randomization involves choosing a list of locations, points in time or individuals, at random and in such a way that all of the possible locations, points in time or individuals in the population have an equal chance of being selected. This can be achieved by using a sequence of random numbers obtained from random number tables (included in some statistics textbooks) or by repeatedly using the random number function on a calculator or computer.

Calculators and computers usually give random numbers between 0 and 1. If you want a series of random integers between say 1 and 36, just multiply each random number by 36, add 1 and take the part before the decimal point. (Occasionally this could give you a value of 37 but you can just ignore this and go on to the next random number.) This system might be used to give a series of random numbers which represent (X,Y) coordinates in metres in a field, numbers of minutes after midnight, or to select individuals from a list where we know how long the list is, e.g. a list of nature reserves where the great crested warbler has been seen (Figure 4.1).

For the sample to be a random selection we must take a *sequence* of successive numbers from random number tables or from a calculator or computer. Suppose we want a sample with 10 members; we should take the first 10 random

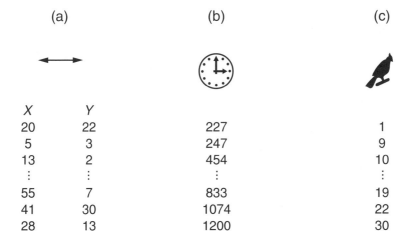

(a)		(b)	(c)
X	*Y*		
20	22	227	1
5	3	247	9
13	2	454	10
⋮	⋮	⋮	⋮
55	7	833	19
41	30	1074	22
28	13	1200	30

Figure 4.1   (a) Random (X, Y) coordinates in metres. (b) Times after midnight (min), selected at random then arranged in order. (c) Selection of nature reserves from a list, selected at random then arranged in order

numbers generated by our computer or calculator, or 10 numbers which follow each other in a random number table. It is permissible to ignore any numbers which *cannot* be used (e.g. because they point to a coordinate outside the sampling area) and go on to the next in the sequence instead. However, we should not try to get a sequence of ascending or descending numbers, or ignore any numbers simply because they are inconvenient or because they make the sample appear a bit unevenly distributed, otherwise the final sample is not a truly random selection.

In the case of an experiment, if we are studying existing populations such as soils in different fields, or grasses from different altitudes, we should use random numbers to decide which individuals to collect or at which points to collect our samples, collect them, and then apply our experimental treatments to these samples. However, in experiments we are sometimes studying populations that don't actually exist anywhere. If we want to know whether a newly developed pesticide is toxic to red ants, our study would aim to look at red ants in general fed with the new pesticide and compare it with red ants in general not fed with the new pesticide.

We *could* do this by treating large areas with the pesticide and then sampling ants from them to see how they have responded, but this is unnecessarily destructive. Instead we would just study samples that represent these populations. The correct procedure in this case would be to collect a sample of red ants large enough to divide up between the experimental treatments and then select half of the ants at *random* to go into one treatment 'with pesticide' and half into the other 'without pesticide', before applying the treatments. In this way we know that the samples we are comparing were the same before we applied the pesticide, so any differences at the end of the experiment must be due to the effects of the pesticide.

Random allocation of a set of similar individuals to experimental treatments is required in many experiments, so I will consider a further example to illustrate why it is important. Suppose you have a set of 60 plants that have been grown in the same conditions on a bench in a greenhouse for 2 weeks and you now want to subject them to four different combinations of nitrogen/potassium (N/P) fertilizer treatments. If you put all the biggest plants in the high N, high P treatment and the smallest plants in the low N, low P treatment, when you come to measure the plants a few weeks later you will have biased the results in favour of finding bigger plants in the treatment with the most fertilizer.

Clearly you would not do this intentionally but it is easy to bias a result by accident. Suppose you start by taking the 15 plants from the front of the bench and giving them high N and high P, and then you take the 15 plants in the next row back and apply high N and low P to them. To the third group of 15 plants you apply low N and high P, and to the 15 plants from the back of the bench you apply the low N, low P fertilizer treatment. The plants might all look the same but those at the front might have had better watering over the last two weeks, or those at the back may be directly over the heating pipes or against the window. We cannot be sure that their roots and other aspects of their physiology are all the same at the start of the experiment.

In a case like this there may be factors that will bias the result that we will never know existed. We can guard against this problem by randomization. Initially we should assign random numbers to each pot (1–60) and then generate a sequence of 15 random numbers for the pots to receive the high N, high P treatment, another 15 random numbers for the pots to receive the high N, low P treatment, a further 15 random numbers for the pots to receive the low N, high P treatment, and the rest of the pots will then receive the low N, low P fertilizer treatment. Every plant will have had an equal chance of going into each of the treatments and the decision about which treatment to apply to each plant has been taken randomly.

If it is *not possible* to use a particular random number (e.g. because that pot has already been allocated), simply ignore it and go on to the next in the sequence generated by your calculator, computer or random number tables. However, numbers should never be discarded simply because they are inconvenient, or because they make things look a bit unbalanced, otherwise this is not a truly random allocation.

The method you use to produce a random sequence of numbers need not be sophisticated. Pulling out numbered pieces of paper from a hat is equally valid.

Random versus systematic sampling

In random sampling we avoid exerting any control over *which* individuals or individual points land in the sample, so we cannot be accused of biasing the sample ourselves. *Systematic sampling* involves using a system, which *we* decide, to determine the individuals or points that will be in the sample. Examples include making measurements at the corners of a 10 m × 10 m grid, or including every 25th item on a list in our sample. This kind of sampling scheme might often seem attractive because we can ensure the sample is spread throughout the population, and there may also be practical advantages to taking samples at regular intervals. However, samples selected systematically rather than randomly have several disadvantages.

Systematic sampling doesn't fit with the theory underlying statistical tests

Statistical tests give probabilities (*P*-values) of getting samples like those observed if the null hypothesis is true (Section 2.5), and the samples were drawn at *random* from the populations. The probabilities calculated therefore allow for the fact that *any* combination of individuals in the populations could land in the samples. If we use a systematic sample (e.g. taking every 10th individual) then not all combinations of individuals can land in the same sample, so the probability calculated in the test may not be accurate.

Sampling points may coincide with underlying processes

Sampling points might coincide with some underlying process that we know nothing about. For example:

- If we are sampling the concentration of a gas in the air and we make measurements every hour, this may coincide with the times when exhaust gases are vented from a nearby factory. Although the vented gases might only affect the surrounding air for 5 min in each hour, they will be present in all of our samples, so we would conclude that the gas concentration was always high at this location.

- When making measurements in a field crop, we may decide to take samples in every 10th row, not realizing that these rows are all directly above field drains and so have slightly better growth than the average for the field.

- While sampling for insects in a field, we decide to make counts at locations on a 20 m × 20 m grid starting 10 m in from the edge, not realizing that 90% of the insects are in the outer 10 m of the field, close to the hedgerows, so we miss them completely.

In any of these cases there would be nothing in our results to tell us that they had, in reality, been biased in this way.

We may subconsciously bias the sample

In any systematic sampling scheme we will have made some decisions ourselves about where or when to sample. We might therefore subconsciously have biased the sample in some way. Even if we have not, it is difficult to prove that we have not, so others will be more sceptical of our results. It is of course possible that even a random selection of points will turn out to be perfectly evenly spaced. However, if we have selected the sample using random numbers, we have not exerted any control over the selection process, so it is clear to everyone that we have not biased the result by our own choices and decisions about which points to include.

What if we can't use random sampling?

For reasons like those given above, many statistics texts argue that *random* sampling is *essential* if we are going to submit the results to statistical testing. Anyone involved in practical science, however, will realize that there are occasions when it is very difficult to select a truly random sample. Take the following two examples.

Example 1: Sampling a chemical in the air

We want to measure the mean concentration of some chemical in the air at a specific place over a day, but it takes us half an hour to make a measurement. We cannot take samples at truly random points in time because we may be required to make two measurements 1 minute apart or less.

Example 2: Crop samples in a field

We are taking samples in a field 400 m × 200 m. The field has a crop in rows 2 m apart (i.e. 200 rows, each 200 m long) and we are not able to cross the rows because they are covered by polythene cloches too large to jump over. If we locate 50 points at random in the field, we might have to walk over 10 km to visit them all, and this would not leave us enough time to make the measurements required at each sample point. There is no point in us repeatedly drawing random samples until we get one where the points are conveniently located, because this would involve us making a decision, so it would no longer be a truly random selection.

If we are unable to use a truly random sampling scheme, we need to think carefully whether to go ahead, and if so how. Remember that if we cannot convince others that our results are genuine effects, we will have wasted our time, even if we have applied a lot of effort to get the results. There are cases, therefore, where we simply have to concede we cannot carry out the study as we would like. However, most users of scientific results appreciate that we do not live in an ideal world and that practical research often involves compromises in some respect.

How might we proceed in the above two examples? In Example 1 we could choose to sample the air at the start of each half-hour, but this is clearly a systematic sample and suffers from the problems described above. A better way would be to choose a sequence of random numbers to represent minutes after midnight and simply ignore any in the sequence that are within half an hour of any of the times already selected. With this scheme, once one time has been selected, we will not include another time within half an hour of it, so this is not a truly random sample; however, any time has a chance of being selected at the outset and we are not exerting any control over which times end up in the sample. This would probably be acceptable to most people.

In Example 2, on crop growth, we might decide we have the time to walk along 10 rows in the day (2 km + the bits at the ends) and still have time to collect the samples at the different points. One option then is to select every 20th row and take points at 40 m intervals along them (total = 50 points). This is also clearly a systematic sample, so we would like to avoid this. A better option is to select 10 rows at random, and select 5 points at random distances along each. It does not involve us in any more work but it is closer to our ideal of a truly random sample. Every point had a chance of being selected at the outset and we have not exerted any control over which points did end up in the sample. Again, this would probably be acceptable to most people.

Unfortunately, I cannot speak for everyone. Whether they consider these compromises reasonable or not is up to them. There is always the risk with samples that are not selected by a truly random process that someone will dispute the validity of our work, so there is a high premium in using truly random samples wherever possible.

4.4   Experimental design

So far in this chapter I have discussed how to choose the number of individuals to include in our samples, and how to use random sampling to select which individuals or points will actually be in the samples. We also need to consider how we will handle the samples and the order in which we will collect them to ensure we are making a fair comparison. Suppose we are comparing the water contents of soil samples from two fields; we might correctly collect a sample from each field, but if we sample one field one day and one the next, we might still bias our results. The term *experimental design* is sometimes used in statistical texts to refer specifically to the physical layout of an experiment or survey, or the order in which measurements are made to ensure that we make a fair comparison. Two of the simplest types of design are described below.

Fully randomized designs

In experiments we are usually aiming to study the influence of only a small number of factors on some property of a population (e.g. effects of fertilizer and water on plant height). However, in practice other aspects of the experimental environment can also affect the results (e.g. light and temperature). Experimental facilities such as greenhouses contain quite a lot of unwanted variation in the environment. While we may not be able to remove this variation, we can at least try to ensure that it affects all of our experimental groups equally. We can do this by assigning individuals from different experimental treatments to positions in our experimental area *randomly*.

This is easier to follow with an example. Suppose my experiment involved growing 6 plants in each of 4 different fertilizer treatments (A, B, C, D) in a greenhouse; I would start with 24 similar plants and first allocate them to the different fertilizer treatments using random numbers (Section 4.3). This is to ensure all the samples are the same before the treatments are applied. However, I also want to ensure variations in the environment in the greenhouse during the trial do not favour some fertilizer treatments more than others. This can be done by placing the 24 pots in a *fully randomized design*. Each plant in each fertilizer treatment is assigned to a position at random (i.e. using random numbers again), as shown in Figure 4.2. If there is any variation in light and temperature it will now affect all treatments to approximately the same degree.

Note that in this example the four fertilizer treatments could have been four different amounts of the same fertilizer, four different fertilizers, or four different combinations of high or low phosphorus with high or low nitrogen. In any of these cases we simply assign a letter to each treatment or treatment combination and randomize their positions in the same way.

Fully randomized designs can also be used where measurements have to be made over a period of time and there is a chance that environmental conditions will change or the individuals themselves will change over that time.

If I wanted to compare the photosynthesis rates of two species of orchid (X and Y) growing in an area of moorland, I would first select a set of randomly located sampling points as in Section 4.3, and take the nearest orchid of the appropriate species at each of them. Locating the orchids may take some time, so I can expect that environmental conditions will change during the time it takes me to make all the measurements. In this case I want to make sure these unwanted variations in environment affect both sets of measurements approxi-

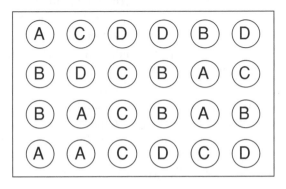

Figure 4.2 Fully randomized design for six replicates of four treatments A, B, C and D

mately equally. Again I can do this using a *fully randomized design*. By making measurements on individuals from the two different species in a random order (e.g. X, X, Y, X, Y, Y, Y, Y, X, X, ...), if there is any variation in conditions while I am making the measurements it will affect the measurements on the two species to approximately the same degree.

Fully randomized designs have the advantage that all treatments will be affected to approximately the same degree by variations in experimental conditions, even if we do not know what variations exist.

Randomized complete block designs

If we know that the experimental environment varies in some systematic way (e.g. we know that one end of a greenhouse is hotter) then we can compensate for this in our experimental design. In effect, we divide our experiment into a number of mini experiments, called *blocks*, and compare the different treatments in each block. The blocks should be chosen so that the environment within each one is fairly uniform. This can be illustrated using the examples from above.

Suppose we knew that the greenhouse we were using for the fertilizer experiment was better lit at one end. We can divide our working area into several blocks along the greenhouse so that one is well lit, the next is slightly less well lit, and so on, until we reach the last block, which is the least well lit. We start by allocating the 24 plants to different fertilizer treatments at random, as before. Next we put one plant from *each* of the four fertilizer treatments in each block. Finally we randomize the order of the different treatments within each block.

Each block has one complete set of fertilizer treatments in it, hence the name *randomized complete blocks* (Figure 4.3). Ideally each block should contain only one plant from each treatment, so in this case we would have 6 blocks, one block for each replicate. However, the experimental area might divide naturally into two areas or three areas, in which case we could have three plants or two plants from each treatment in each block. It is important that each block should have the same number of plants from each of the treatments.

In the case of the orchid survey, we might realize that light conditions are likely to change systematically over the day, so we can improve the experimental design by treating periods of time as blocks and ensuring that we measure one of each species in each period of time. The length of the periods will depend on how long the measurements take (e.g. perhaps in each 45min we can measure one of each species). The order in which we take measurements from the different species in each 45 min period should then be randomized; we should not always start with species X, for instance. The survey would then be said to have a *randomized complete blocks* design.

Blocks can be any factors that we think may influence the results, but we are

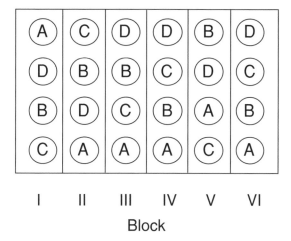

Figure 4.3 Randomized complete blocks design for six replicates of four treatments A, B, C and D

not interested in what their effect actually is. We might carry out identical mini experiments in several different plant growth cabinets, or with several different families of animals, or on several different measurement dates, or using several different seedlots. Then plant growth cabinet, family, measurement date or seedlot would be treated as a blocking factor in the analysis, just as areas of the greenhouse or period of the day were treated as blocks in the earlier examples. We should randomize the order in which the different treatments are arranged within each block. These would all then be randomized complete block designs.

Further points about experimental designs

Randomization in experimental designs is carried out for the same reason as randomization during sampling, i.e. in order to avoid bias from unwanted and possibly unknown influences. The randomized complete blocks design helps to ensure that foreseeable but unwanted influences affect all experimental treatments equally. An important point to note here is that analysis of both fully randomized and randomized complete blocks designs is very straightforward. In fact, for the fully randomized design we do not need to make any adjustments to the statistical tests at all.

If we have used a randomized complete blocks design, we do not *have* to make any adjustments to the statistical analysis either. Simply laying out the trial in this way is useful because it helps to ensure all of our experimental treatments or survey populations are influenced to the same extent by unwanted variations in the experimental environment. However, randomized complete blocks designs have the advantage that we can choose to take account of the blocks during the statistical analysis, which will improve our chances of finding differences between treatments if they exist. Section 8.4 explains how to do this.

4.5 Further reading

There are many other types of experimental design that have been devised for specific situations, particularly agricultural trials and to improve manufacturing processes, some of which might be useful in environmental or biological studies. I do not cover them in this textbook. Unlike the two designs given here, their analysis can become very complicated, so before carrying out the experiment, I advise anybody attempting to use them to ensure they know how to analyse the results and to obtain a suitable software package to do the calculations. There are also pitfalls in that unlike fully randomized and randomized complete blocks designs, these other designs do not always produce equally powerful tests of all the factors or combinations of factors included in the experiment. It is often better to try rethinking your experiment or survey so that you can use a fully randomized or randomized complete blocks design.

Balanced incomplete blocks

Balanced incomplete blocks are used when it is not possible to include all of the treatments in every block; perhaps we had five sites but it was only possible to visit four in a day.

Nested designs

Nested designs are used when different sets of treatments are used in different parts of a trial. For example, we may compare the growth performance of five different species of eucalyptus in four different countries, each with three concentrations of potassium fertilizer. However, it is not possible to obtain exactly the same type of fertilizer in each country, so we can only genuinely compare the effects of fertilizer concentration *within* each country. The fertilizer treatment is said to be nested within country.

Split plots

Split plots are a particular type of nested design, used when it is not possible to lay out a trial in a fully random way for practical reasons. For example, to compare different variety × irrigation treatments in a field, ideally we would like to randomize the location of each variety × irrigation combination but it would not be practical to have different irrigations on adjacent small areas. Instead the irrigation treatments may be applied to several *main plots* and each main plot is split into several *subplots* containing the different varieties.

Response surface designs

Response surface designs are used to identify the optimum amounts of various treatments in combination to achieve some effect, e.g. to identify the best combinations of chemical and biological additives to clean up an oil spill.

5

Exploratory Data Analysis and Data Presentation

5.1 Introduction

Using graphs and tables to present data is an integral part of most studies. However, we often leave this to the very end, when we are clear about exactly what we want to present. Having carried out a study with a particular statistical test in mind, we are keen to press on with the statistical analysis and 'get that P-value'. If the statistical test supports our theory, a thousand feathers tickling the soles of our feet would not persuade us to give up our result. We hurriedly write it up – thus my theory is supreme – and look no further at the data.

In reality, we should use graphs as the first stage of data analysis, before using a statistical test, as well as at the end to present our findings. There are several good reasons for looking at our data graphically before using statistical tests:

Graphs help to verify that it is valid to use a particular test

Statistical tests are only valid if certain conditions are met by the data (Chapter 6). If the conditions are not met, we will not be aware of this from carrying out the test, but the results of the test will be misleading us, i.e. wrong! In particular, the distributions of the values are often assumed to be Normal and the variances to be homogeneous. Don't worry about what these mean just now; they are explained in Chapter 6. But checking for these conditions will involve graphing our experimental data.

We would plan to use a particular type of test based on what we *expect* the data to be like, but we should still *check* once we have collected the data that the distributions of values in the samples are consistent with these expectations. Unfortunately, it probably will cause us more work if our expectations turn out to have been wrong, but what is the point of going ahead with a test that is (or even might be) giving us misleading answers?

Graphs may reveal unexpected patterns in the data

Suppose I have a theory that my new antipollution treatment for rivers, Whoosh-clean, will increase the numbers of invertebrates. I set up a suitable trial to compare the effects of adding 0, 1, 2 or $3\,\mathrm{mg}\,l^{-1}$ of my chemical to a number of rivers. The numbers of invertebrates I get in my samples average out at 100, 120, 140 and 2500 respectively. I also know from other people's studies that I would expect to get around 180 in samples collected in the same way from a 'clean' river.

If I simply carry out a statistical test, it will almost certainly support my theory that Whoosh-clean increases invertebrate numbers. However, if I looked at the data first, I would see that something unexpected has also happened (numbers are very high in the $3\,\mathrm{mg}\,l^{-1}$ treatment), which should set me wondering what the chemical is really doing. One explanation is that my new chemical has actually killed something that feeds on the invertebrates, i.e. Whoosh-clean is itself a pollutant! The important point here is that *statistical tests* give answers to very specific questions, e.g.

Does adding Whoosh-clean lead to an increase in invertebrates?

They will not alert us to the fact that there are other important messages in the data as well. Graphical presentation of the data is usually the best way to find this out.

Graphs quickly reveal any mistakes in our data

We might have made a mistake entering one of the values; perhaps we have entered 105 instead of 0.105. The statistical test will just assume we really measured this, but if we looked at our data graphically, we would quickly see that a mistake had been made.

Studying our data using graphs or other means simply to get a good impression of what is going on and whether anything unexpected has happened is called *exploratory data analysis*. Whether we are using graphs for exploratory data analysis or to present our findings, a number of types of graph are commonly used and these are described below. For the sake of clarity I have left an explanation of error bars until Section 5.7. However, in reality, if a column or line graph is used to present results, error bars should almost always be included as well.

5.2   Column graphs

Column graphs, sometimes called bar graphs, are probably the most common way to present data in scientific studies. An example is shown in Figure 5.1.

The height of each column represents the mean value found in the sample. However, column graphs can be used to show medians or some other value instead, so it should be made clear in the figure caption exactly what the graph shows in each case.

The scale starts at zero and the bars are all the same width. Subconsciously we tend to see the overall size (i.e. area) of the bars as indicating the relative magnitudes of the things they represent, so in this case we are gaining an accurate impression of the real differences between roads. If the scale were to start at 5, a value of 11 would produce a bar six times as tall as a value of 6. Although we could still read the values correctly off the graph, we might gain an impression that there was a very large difference between these two, when the difference is really less than double.

This effect is sometimes used in the media and by pressure groups to make things seem more important than they are. Since our aim in science is to guide people towards a correct understanding of something, whether or not it supports our theory, we actually have to go out of our way to avoid giving false impressions. On rare occasions when you might decide that you have to start the axis at some value other than zero – perhaps because the bars appear almost the same length but the small differences between them have practical importance – you should point this out to the reader.

When you have been studying the effects of more than one factor, it is often

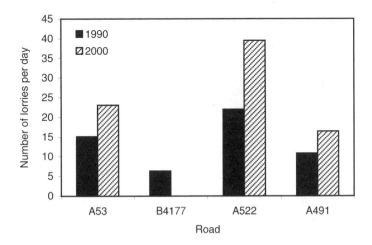

Figure 5.1   Mean numbers of lorries per day leaving Grimley quarry along main routes out of the area

useful to try grouping the data in different ways to reveal the effects of the different factors. Figure 5.2 shows the effects of different combinations of weedkiller *and* soil type *and* growing clover during the period between crops, on the nitrogen content of plants. The same data are presented in all three graphs.

Note how the first figure emphasizes the effect of weedkiller, the second emphasizes the effect of soil type, and the third emphasizes the effect of growing clover during the period between crops. We would not usually present all three

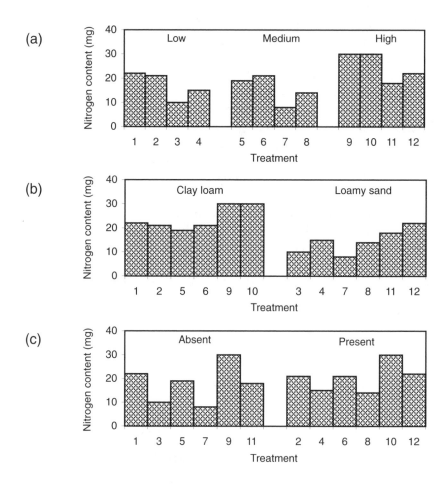

Figure 5.2   Mean nitrogen contents of plants in different experimental treatments: (a) grouped by weedkiller treatment, (b) grouped by soil type and (c) grouped by presence or absence of clover during period between crops. Treatments are a combination of three factors: weedkiller, soil type and clover during period between crops. Weedkiller: low (treatments 1, 2, 3, 4), medium (5, 6, 7, 8) and high (9, 10, 11, 12). Soil type: clay loam (1, 2, 5, 6, 9, 10), loamy sand (3, 4, 7, 8, 11, 12). Clover during period between crops: absent (1, 3, 5, 7, 9, 11), present (2, 4, 6, 8, 10, 12)

of these graphs in a report, but it is useful to try out different orderings like this, particularly during the initial exploratory data analysis.

You may also see graphs similar to this being called *histograms*. In a column graph, values are represented by the *height* of the columns; in a histogram, values are represented by the *area* of the columns. If the columns are all the same width, this amounts to the same thing. However, in a histogram the columns are not always of equal width. Histograms are mostly associated with showing frequency distributions. For example, we might have counts of the numbers of values in each of the ranges 0–5, 5–10, 10–20, 20–30 and 30–100. These would be shown as columns of width 5, 5, 10, 10 and 70 respectively, and the *area* of each column would be in proportion to the total number of values in that range.

5.3   Line graphs

In some situations it is reasonable to use either a column graph or a line graph, but the choice of which to use is usually based on whether there is some natural order to successive points on the graph (e.g. successive points in time or successive points along a river). In such cases, line graphs have the advantage that they show clearly which points follow on from which others. Figure 5.3 shows an example of a typical line graph.

Notice how the lines connect points in a logical sequence; in this case they are successive points in time. Strictly speaking, a line graph makes no claims other than the actual data points shown, but there is often an implication in line graphs that, although we have only measured at certain points, the values

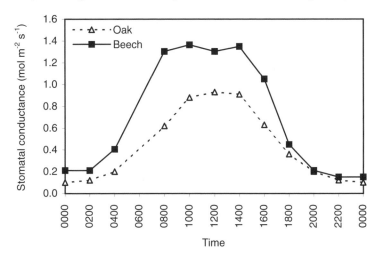

Figure 5.3   Changes in mean stomatal conductance of two tree species over the course of a day

probably changed gradually from one to the next – as suggested by a sloping line – rather than a step change in values.

If the points are spaced along the horizontal axis with equal spacings for equal time intervals or distances, how steeply the line slopes between one point and the next also helps to create an impression of how quickly the values are changing at different stages along the line. This means that line graphs give a particularly good representation of the data if the values do tend to change gradually between one point and the next.

Conversely, if joining successive points by a line creates a misleading impression, we should avoid using a line graph. Suppose we measure soil clay content at 500 mile intervals around the coast of South America; successive values really tell us nothing about the soil in between. We would be better to use a column graph in this case. Because the measurement at 6:00 am was missed for some reason, a gap has been left to maintain equal spacings between times, otherwise the reader might gain the impression that there was a more rapid change between the 4:00 am and 8:00 am points.

If we have good reason to believe the values do change predictably between one point and the next, due to some underlying process – for instance, concentrations of pollutants are likely to change fairly steadily with distance in water – we might choose to join the points with a curved line (Figure 5.4). Again, our aim in how we present the graph should always be to help the reader gain an accurate interpretation of the facts, not simply to make them believe our theory is correct.

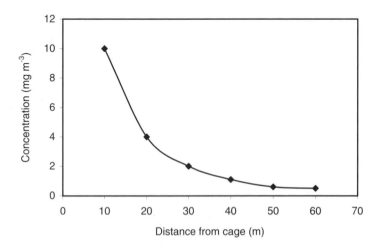

Figure 5.4   Mean concentrations of insecticide at different distances from fish cages at the Inverslochit fish farm

5.4   Scatter graphs

Scatter graphs are used to show the relationship between two variables (e.g. length and area) measured on the same group of individuals. An example of a scatter graph is shown in Figure 5.5.

Each point represents an individual for which two characteristics have been measured; here the chararcteristics are mean mass per seed, and number of seeds produced. The overall collection of points gives us a picture that is not apparent from any individual measurement. In this case the points seem to be clustered around a straight line. Using a correlation test (Chapter 9), we can test whether this is just a chance event in the sample or whether this is likely to come from an underlying straight line relationship in the population.

It is common to include the results of a correlation analysis on the graph itself or in the caption; r describes how closely the measured points are scattered round the line, and the P-value tells us whether we can conclude there is a straight line relationship in the population that the sample came from. Both r and P therefore give important information. Correlation and the meaning of r and P are described in Chapter 9.

Another thing that we can see from Figure 5.5 is that one point (circled) does not seem to fit in with the trend. We can consider what this point represents. In this case it is a plant that produced more seeds than we would expect for seeds of that size, or looking at it another way, bigger seeds than we would expect for a plant producing that many seeds.

We could identify from our dataset which plant species this was, and this

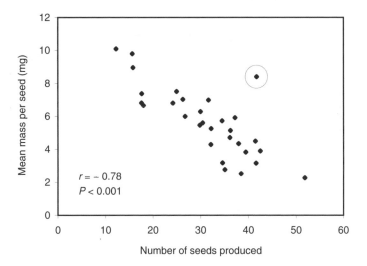

Figure 5.5   Relationship between the mean mass per seed and the number of seeds produced, for a range of plant specimens collected on Mossely Heath. The circled point appears to be an unusual observation, which could warrant further investigation

might turn out to be the most interesting find of our study. Notice that we could not appreciate that this species was unusual from observing it on its own, nor would it be apparent from a test for correlation. However, when compared to a range of other plants on a graph, this point clearly stands out. This is why exploratory data analysis is such an important component of analysing experimental or survey data.

In some cases we are able to determine that one of the measured characteristics is controlling the other; how fast individual lions can run might be controlled by how long their legs are, but the length of their legs is not controlled by how fast they run. If we can determine this, then the controlling variable should be plotted on the horizontal axis, or the X-axis. We would therefore plot leg length – called the control variable or the independent variable – on the X-axis, and running speed – called the response variable or the dependent variable – on the vertical axis, or the Y-axis.

There is no particular advantage in plotting them this way round except that it is a widely used convention, so people tend to interpret graphs assuming the variable on the X-axis to be controlling the variable on the Y-axis. If neither variable can be considered to be controlling the other, it does not matter which way round the graph is plotted.

In cases where the measurement on the Y-axis is dependent on the measurement on the X-axis, it is possible to produce an equation to predict Y-values from X-values using a technique called regression. This will be explained in Chapter 9. The equation and the line that it describes are then also commonly shown on the graph (Figure 5.6). As you can see from this example, relationships are not always straight lines.

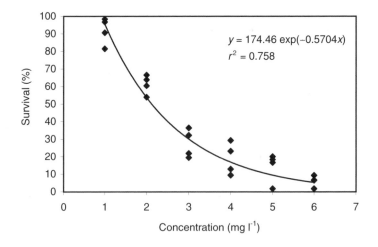

Figure 5.6   Effects of different concentrations of an unidentified substance on survival of *Pseudomonas fluorescens*. Note, the equation of the fitted line is shown, where y is the percentage survival and x is the concentration in mg l^{-1}

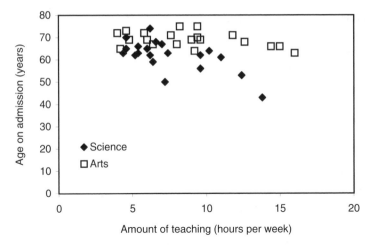

Figure 5.7 The relationship between age on admission and number of hours per week of teaching in their final year of work, for university lecturers from science and arts subjects admitted to psychiatric hospitals

It is also possible to represent more than one population on the same graph by using different symbols (Figure 5.7). In this case it appears that different straight line relationships exist in the two populations. Chapter 9 describes how to carry out a statistical test for this.

5.5 General points about graphs

Printed graphs usually have a figure caption

In printed works, the caption forms part of the graph. Most books and journals adopt the convention that figure captions appear immediately below the figure. The caption should be distinct from the rest of the text on the page and this is often achieved by using a smaller font, setting it in italics, or by centring it across a narrower text width. A rule of thumb is that if you were to take the figure and its caption away from the rest of the text, there should be sufficient information for a reader to make sense of it without having to refer to the rest of the text. Clearly there is a limit to how much detail you can provide in a figure caption, but

Mean trout lengths in five different ponds in northern England.

is a better caption than

Fish lengths (see text).

Graphs for scientific articles seldom have a title; this information usually goes in the caption.

Slideshow graphs need a title but minimum labels

Graphs for projecting during talks should have very few words on them and not very much data. Captions are generally omitted as they would be too small for an audience to read, but it is useful to include a title on the graph. The talk itself should also make it clear what data are being presented. You will see plenty of examples of what does and does not work well by attending talks. The graph should include a description of what is shown on each axis (often just one word) and the units. If data for more than one treatment or population is shown, the graph should also include a key explaining what each set of values represents.

Examples of how the same data might be graphed for a report and a talk are given in Figure 5.8. Error bars, the small projecting lines, are explained in Section 5.7. Given time to study the graph, all the data in Figure 5.8(a) can be understood. However, this is difficult in a short time, and the large number of lines will be difficult to see in a large room. Figure 5.8(b) shows a selection of the data but clearly presents the main messages, that chloride tracer concentration decreased with distance from the source and lower concentrations were found in areas with deeper peat.

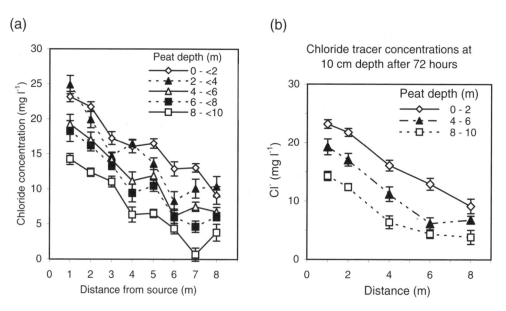

Figure 5.8 Alternative ways of presenting data from a study of chloride tracer concentrations at 10 cm depth in peats of different depths, at different distances from the source after 72 hours: (a) a possible format for a written report, (b) a possible format for a talk

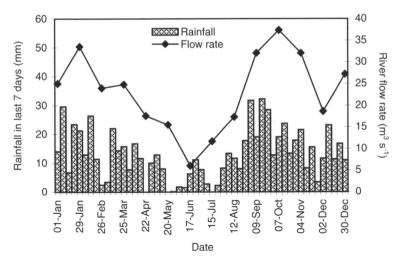

Figure 5.9   Weekly rainfall and flow rate measured at four-week intervals at the lower Stems Ghyll gauging station

Scatter graphs may sacrifice information to show relationships

If I wanted to show how rainfall in an area related to river flow, I could use a scatter graph. An alternative is to use a separate axis on each side of the graph (Figure 5.9). This shows that the two properties tend to vary together and also indicates when the peaks occurred. On a scatter graph we would lose the information about when the peaks occurred but we would get a clearer view of how closely rainfall and river flow were related.

Choose the most effective graph for what you want to show

Many other types of graph can be produced easily using computers, including 3D column graphs, surface plots, pie charts and contour plots. There are disadvantages to these in that it is often difficult to read values from them accurately and difficult to present error bars (Section 5.7) or some other measure of the 'margin of error' on them. That is why column graphs, line graphs and scatter graphs are the most widely used. Ultimately, though, the purpose of a graph is to be informative. For presenting patterns and trends in data, one of these other graphs may be the most effective.

5.6   Tables

When presenting data, tables can often be used as an alternative to graphs. Here are some situations where a table might be better:

- When it is important that the reader can obtain the values in your results accurately or if readers are likely to want to use or compare your actual figures elsewhere, rather than just to compare with other populations in your experiment or survey.

- When you need to present a lot of data and this appears messy when you try to draw a graph of them.

- When data can be presented a lot more compactly in a table (in printed works such as journals, space costs money).

It is common practice, though not universal, to place the caption *above* the table. As for figure captions, a table caption should give enough information for the table to make sense without referring to the text.

5.7   Standard errors and error bars

Experienced readers of scientific literature will have noticed straightaway that I have missed off the error bars from most of the graphs so far. This was only because I wanted to introduce them here. As explained in Section 2.3 and elsewhere, experimental and survey results are virtually always measurements or observations on a *sample* of individuals, whereas the subject of our study is really the *population* it came from.

On the graph we would usually present the mean of the *sample*. So far as the mean of the *population* goes, we can only present a *range* of values in which it probably lies (Section 2.4). On graphs this is usually achieved using error bars, as in Figure 5.10. We should try to get used to thinking of graphs as presenting the range of values where the mean probably lies, rather than exact values. The true population mean is probably somewhere in the range between the upper and lower ends of the error bars. On column graphs it is usual to show just the error bar projecting upwards from the sample mean, though there is really the same range of likely values for the population mean below the sample mean as well.

Given that the real information about the populations being studied is a range of values where the mean probably lies, rather than the height of the column or position of the line itself, there are very few times where it is appropriate to present graphs without error bars or some equivalent indication of the 'margin of error'.

Error bars shown on graphs usually represent the standard error (Section 2.4) but they do not have to. The range $\pm$ one standard error is the range in which we can be approximately 68% confident that the population mean lies. However, error bars shown on graphs can have other meanings too (e.g. they may be used to represent the 95% confidence interval).

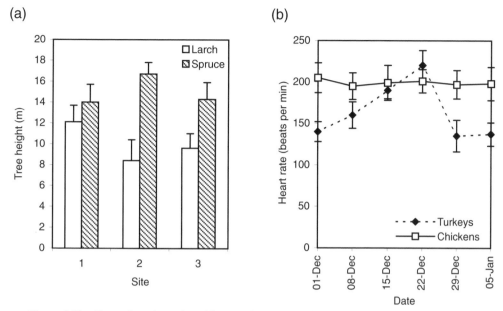

Figure 5.10   Examples of graphs with error bars: (a) a column graph, (b) a line graph. The caption to a graph should make it clear whether the error bars represent the standard error or 95% confidence interval, or some other measure of error

If the heights of the columns represent sample *medians*, rather than the sample means (Section 13.1), it is common to show the *interquartile range* in the form of error bars. This is the range which cuts off the top 25% and the bottom 25% of values in the sample, i.e. the range containing the middle 50% of values in the sample. Because error bars can be used to represent a number of different things, it is important to make it clear in the caption to a graph what the bars represent in that case, e.g.

Figure 4. Mean NO_2 concentrations at the three sites (± 1 s.e.)

If the data are presented in a table, standard errors (or the 95% CI) should also be presented. Table 5.1 shows a typical method for doing this. The number in parentheses after each mean represents the standard error for that mean.

Table 5.1   Effects of atmospheric CO_2 concentration and pot size on mean heights of tree seedlings after one year (standard errors in parentheses)

CO_2 concentration (ppm)	Mean height of seedling (mm) in pots of volume		
	$0.2\,dm^3$	$0.6\,dm^3$	$2.0\,dm^3$
350	222 (15)	365 (21)	542 (18)
600	228 (12)	370 (15)	565 (20)

6

Common Assumptions or Requirements of Data for Statistical Tests

6.1 Introduction

In order to use statistical tests on our data, certain criteria have to be met. This chapter begins by looking at why tests have these kinds of restrictions on their use, and then at how we can decide whether our data meet the requirements. It concludes with Section 6.3 on transforming data, which considers how we can sometimes overcome the problem of our data not meeting these requirements.

Why do statistical tests need assumptions?

Suppose I select a random sample of six trees from an even-aged, unthinned plantation of Sitka spruce. I measure their heights in metres: 14.2, 12.9, 12.7, 16.0, 15.1 and 14.4. I can calculate the sample mean (14.2 m), and the chances are this is a fair estimate of the mean height of trees in this plantation (the population mean). I can calculate the standard deviation (1.27 m) and the chances are this is a fair estimate of the standard deviation of heights of trees in this plantation. I could therefore use these figures to estimate what the heights of the rest of the trees in the plantation are likely to be. This will involve making certain assumptions.

If I *assume* that the population has a Normal distribution, the distribution of tree heights in the plantation probably looks something like Figure 6.1(a). This is a Normal distribution with mean 14.2 and standard deviation 1.27. If I *assume* the population has a Uniform distribution (i.e. with heights spread evenly over a range of values), its distribution would probably look something like Figure

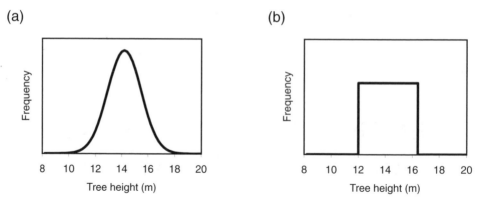

Figure 6.1   (a) Normal distribution with mean 14.2 m and standard deviation 1.27 m. (b) Uniform distribution with mean 14.2 m and standard deviation 1.27 m

6.1(b). This is a Uniform distribution with mean 14.2 and standard deviation 1.27. What we *assume* about the *shape* of the distribution of heights affects what we think the heights of the other trees in the plantation are likely to be.

Suppose I wanted to decide whether the tree heights in another plantation had the same distribution. I would of course have to take a sample from the other plantation and compare it with the first sample. If the populations had the same distribution, I would expect the samples to have similar means and standard deviations. If both populations had identical Normal distributions, I would also expect to find most of the values in the second sample to be around 14.2, with some higher and lower values (Figure 6.1(a)).

If both populations had identical Uniform distributions, the values in the second sample would probably be spread fairly evenly over the range 12.0 to 16.4 but none would be outside this range (Figure 6.1(b)). Therefore, if I asked you to judge whether another sample came from a population with the same distribution as the first, it would be helpful to you to know what *shape* the distributions were thought to be. This information would help you reach your verdict.

In a statistical test we are often asking the question, how likely is it that different samples have come from populations with identical distributions? Calculating the means and standard deviations of the samples will help, but as we have seen, the decision will be easier if we assume something about the *shapes* of the distributions. If we give a statistical test some information about the shape of the distribution, as well as the sample data, it will also be able to give us a clearer indication as to whether different samples come from identical populations or not.

In some sophisticated statistical tests you can actually decide what shape of distribution to use as one of the inputs. Most tests, though, have fixed assumptions. If it is fair to assume that the populations have the required shape of distribution, you can use the test; if not, you can't. The assumptions of individual tests are described in the relevant chapters of this book. Many tests have been produced based on the assumption that the populations have Normal distributions because this shape of distribution occurs naturally in many situations. However, not everything we might want to measure will have a Normal distribution, so it is important to check our data if we want to use one of these tests.

How should we determine the shape of distribution?

Most statistical tests do not check this as part of the test. It is assumed that if you have decided to use a particular test then *you* have already checked that the shape of the distribution is suitable. We have already seen that the sample of tree heights could have come from a Uniform distribution or a Normal distribution, or even some other shape of distribution. If we only have the measurements in the sample to go on, how should we decide which is correct? There are two ways.

Look at the distributions of values in the samples themselves

Looking at the distributions of values in the samples only really works if we have a reasonably large sample size. With so few values, the distribution in Figure 6.2 is not very informative. If we have samples with rather more values (e.g. > 30) then a clear shape will emerge. This will be roughly the same shape as the population from which the samples came. There are also statistical tests that can show whether a sample is unlikely to have come from a particular shape of distribution (Section 6.2).

Biological knowledge or experience of similar situations

Biological knowledge or experience of similar situations would tell us that a Normal distribution is a fair assumption in the case of the unthinned tree plantation. There is no reason why we would get quite a lot of trees growing to 16.4 m but none larger. It is much more likely that most trees will grow to a similar height with a few unusually large and a few unusually small. Note that if the plantation had been thinned (e.g. the smallest trees had been selectively cut down) the heights of the remaining trees would probably not be Normally distributed.

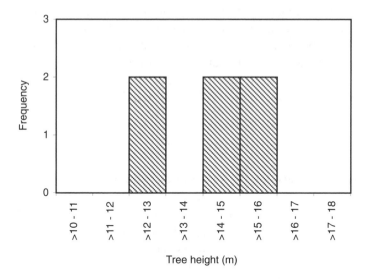

Figure 6.2   Frequency distribution for the six tree heights in the sample referred to on page 77

Parametric tests and non-parametric tests

Parametric tests involve fairly restrictive assumptions, usually that the populations being compared have Normal distributions and equal variances. They include *t*-tests, ANOVA and regression. However, because these conditions are often met by the kinds of populations studied in environmental and biological sciences, they are the most frequently used.

Non-parametric tests, also known as distribution-free tests, have been devised to cope with situations where the populations are not Normally distributed, or where it is not known. Despite the name, most of them have some assumptions about the shape of the distribution, but they are less stringent, e.g. one test requires that the distributions must be the same shape but not necessarily Normal.

It might seem sensible to use non-parametric tests all the time to avoid the risk of making an incorrect assumption. However, just as you or I would make use of the information that two samples came from Normal distributions, when comparing them, so does a parametric test. Non-parametric tests have only the values in the samples themselves to go on and generally produce less conclusive results. Parametric tests are therefore more powerful than their non-parametric equivalents and should be used whenever the assumptions of the test are valid.

6.2   Common assumptions

This section describes some of the most common assumptions in statistical tests and ways of checking whether they are met. Which of these are required by particular tests is indicated in the descriptions of the individual tests in Chapters 7 to 15.

Independent random samples

These are assumptions about the way you have collected the data. In effect, the tests assume you have collected samples that are truly representative of the population. These assumptions apply to all of the statistical methods in this book.

Independent measurements or observations

It is assumed that each individual in the sample would behave in the same way, regardless of how the other members of the sample behave. If we want to know how long it takes an average sheep to find its way out of a maze, we might choose to study 10 sheep, i.e. 10 replicates. We would need to test each sheep on its own because if we put them all in together they will behave as a flock. In that case, if any one of the 10 found its way out, the rest would probably all follow. Statistical tests assume that no member of a sample is influenced in its behaviour by the others. If we want to study the behaviour of sheep in flocks, we need to have independent replicate flocks, not treat the individual sheep within a flock as replicates.

Random sampling

The individuals or individual points in a sample should be selected by some random process (e.g. a series of random numbers from a computer) in such a way that every individual or point in the population has an equal chance of being selected. If we do this, we *might* get mostly unusually large or unusually small values in the sample. The tests assume samples have been selected in this way and the probabilities given by the tests allow for this. Sections 4.3 and 4.4 contain more details about random sampling, and randomization in experimental designs.

Normal distributions

The Normal distributions assumption relates to the distributions of the populations being studied, not the samples themselves. For us to accept this assumption, it must be reasonable on theoretical grounds, i.e. we must expect that
values will be concentrated symmetrically round some mean value, and any
previous research should not contradict this.

The distribution of values in the *samples* should also appear approximately
Normal. Many computer packages will draw a frequency distribution for you
(the function is sometimes called Histogram). Otherwise you could do the same
by hand. You need to divide up the range of values into categories. Dividing it
into $\sqrt{n}$ categories is usually quite effective for showing the shape well, where n
is the number of values in the sample. For example, if you have 45 values, divide
the range between the highest and lowest into 6 or 7 equally spaced categories
and count the number of values which fall in each category. Distributions like
those in Figure 6.3(a) and (b) require transforming (Section 6.3). Something
close to Figure 6.3(c) does not.

Some computer packages have the option of producing a Normal plot (or a
plot of N-scores). This is an alternative to plotting a histogram of the frequency
distribution. If the values have a Normal distribution, they will fall on a straight
line or nearly so. The distributions in Figure 6.3 are shown as Normal plots in
Figure 6.4.

The decision as to how close to Normal the distribution needs to be is a
contentious one. To give an objective decision some workers use a statistical
test (such as the Anderson–Darling, Ryan–Joiner, Kolmogorov–Smirnov or
chi-square test). A *P*-value of ≤ 0.05 signifies the sample probably does *not*
come from a population with a Normal distribution. Conversely, a *P*-value of
> 0.05 signifies that the sample quite conceivably *does* come from a Normal
distribution; this is usually taken to mean that no transformation is required.
An advantage of this is that it gives you a statistic with which to argue your case

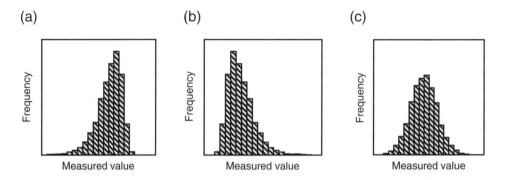

(a)   (b)   (c)

Figure 6.3   Frequency distributions: (a) skewed to the left, (b) skewed to the right, (c) Normal

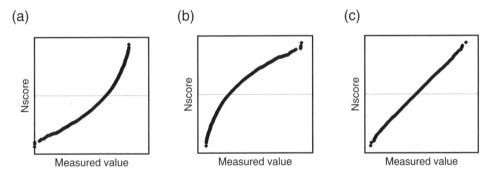

Figure 6.4   Normal plots equivalent to the frequency distributions shown in Figure 6.3: (a) skewed to the left, (b) skewed to the right, (c) Normal. Normal plots may also have the vertical axis labelled 'normal score' or 'probability'. If the points do not lie approximately along a straight line, this implies the population does not have a Normal distribution

for accepting or rejecting that the data come from a Normal distribution.

Statistical tests for Normality are not ideal because the tests are very insensitive to non-Normality for very small samples, and perhaps too sensitive for very large samples. A visual examination of a frequency distribution for signs of skewness (non-symmetry) or extreme values, both of which indicate Non-Normality, is often a more useful check. Many statistical programs will generate random data with a Normal distribution and draw a histogram of the values. It can be useful to try this several times using the same sample size as your experimental data to get your eye in as to what samples from a Normal distribution look like.

In a straight comparison of two population means using a t-test, it is simplest to plot the frequency distributions for the two samples separately. They should both be approximately Normal. In ANOVA it is the distributions of values *within* each of the populations we are comparing, and in regression the distribution of values at any point along a regression line, which should be Normal, not the full set of original measurements. The easiest way to check this is to carry out the ANOVA or regression and choose to save the *residuals*; most computer packages will do this. Then plot a frequency distribution of the residuals. If *this* distribution is Normal, you can go ahead and use the result of the test. Otherwise the data should be transformed (Section 6.3) and the test run again.

Equal variance

Equal variance is also sometimes referred to as homogeneous variance, stable variance, constant variance, or homoscedasticity. What it means is that to give accurate results, statistical tests often require the 'spread', technically the variance, of individual values to be the same in each of the populations we are

(a)   (b)

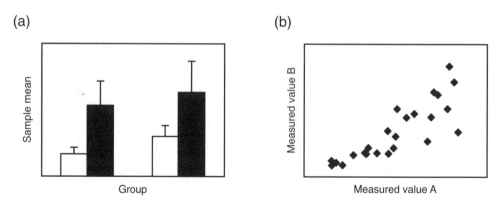

Figure 6.5   Situations likely to require data transformation: (a) larger means have larger error bars; (b) scatter is greater at the higher end of the line

comparing. As for the assumption of Normality, if we are to accept this assumption it must be reasonable on theoretical grounds and not contradicted by the data in the samples. Unequal variance occurs quite commonly because groups with high values tend to have more spread in their values than groups with low values. For example unfertilized plants might range in height from 10 to 15 cm, whereas in a fertilized treatment most are between 20 and 30 cm. Not only is the mean greater in the fertilized treatment, but also the spread of values. Some practical ways to detect this are shown in Figure 6.5.

To compare just two variances the F-test can be used (Chapter 7). Bartlett's test can be used to give an objective check for homogeneity of variance where more than two populations are involved (e.g. when using ANOVA); some computer packages have this. A P-value of ≤ 0.05 can be taken to mean the populations probably do *not* have equal variance. However, like the tests for Normality, it is an insensitive test if the sample sizes are small and perhaps too sensitive if the sample sizes are large. Examination of graphs for obvious systematic differences, like those in Figure 6.5, is often the most useful method.

Problems with non-homogeneity of variance are often cleared up in the process of transforming data to bring about a Normal distribution. So if you found a need to correct for non-Normality, do this *before* you check for homogeneity of variance in the transformed values. If you need to correct specifically for non-homogeneity of variance, the log transformation is a good one to try first (see below).

6.3   Transforming data

The biggest problem people have with transforming data is that the theory behind it is complicated and, at times, counterintuitive. However, the practicalities of transforming data are straightforward, so if we are equipped with a basic understanding of what it means, when to use it, and how to do it, it will not present us any problems.

What is data transformation?

This is a process in which *each* of the measured or observed values is 'converted' in some way to produce a new set of values, and the statistical analysis is then carried out on these new values. Bear with me, I will explain why in a moment. There are many possible ways to 'convert', i.e. transform, values but those in common use are usually something simple like taking the square root of each value. A key feature of transformation is that we do the same thing to *each* value. Therefore *each* of the individual measurements or observations in our sample produces an individual transformed value.

Why is data transformation useful?

As discussed in Section 6.2, many statistical tests have assumptions that the populations we are studying have Normal distributions and/or equal variances. If, for instance, we have counted the numbers of insects on leaves and our check shows that these are not Normally distributed, we could not use a *t*-test to compare the mean number of insects per leaf in different treatments. Supposing we changed our minds and decided to compare the means of log(number of insects per leaf) in the different treatments, i.e. for each leaf in the sample we record the logarithm of the number of insects on it, rather than the number of insects itself. If the distributions of log(number of insects per leaf) turned out to be Normal, we *could* use a *t*-test to compare them. It may be helpful therefore to think of transformation not so much as changing the values, but more as changing our minds about which values we are going to study.

Why then, if we are interested in whether *counts* of insects are greater on one type of leaf than another, might it be acceptable, or even better, to analyse *logs of counts* of insects instead? For a long time I got by telling students, 'You just have to accept it.' Indeed this can be quite an effective approach. However, as scientists you probably have enquiring minds and would like something a little better. I am indebted to one of my students for teaching me something about this subject on a day when I tried to appear cleverer than I really was. The student asked me if I could help her choose a transformation to use on her data. I was busy and thought I would buy time by saying, 'I will, when you can explain to me why you think transformation would be useful here.' I expected her to come back some time later. 'Okay,' she said, 'it's like this.' And, with a little adaptation, it's like this.

Data transformation is most often used when we plan to use a statistical method that compares *means*, or uses means in some other way (e.g. *t*-tests, ANOVA and regression). We have to remember here that our real aim is probably to compare 'typical' values in the populations, and calculating the *mean* often happens to be a mathematically convenient and objective way to say what a 'typical' value is. This is all very well, but there are instances when the mean of a set of numbers is anything but typical. Take the following sets of values. These might be, for instance, distances flown in metres by garden birds between feeds, in two different suburbs in winter.

| Sample A | 30, 49, 74, 40, 63, 295, 60 | mean = 87.3 |
| Sample B | 54, 42, 96, 58, 65, 98, 88 | mean = 71.6 |

The first set has a mean higher than the second, although typical values in the two sets do not appear very different; 87.3 is certainly not typical of the distances flown in sample A. The mean of sample A is greatly influenced by the value 295. We do not want to drop any values, because they are part of the dataset (in this case one bird did fly 295 m to its next feed). And where would we stop? Would we also drop the 74 value? In fact, dropping correctly measured values from a dataset is equivalent to making up your data. In effect, we would be choosing the mean value ourselves, so we might as well make up the rest of the data as well. Of course, then it is no longer a scientific study.

What we need is an objective method that will *reduce the influence* of any atypical values so that when we use a statistical test we will indeed be comparing typical members of the populations, and not just how a few unusual individuals behaved. We achieve this by *transforming* the values. I will describe shortly how to choose the type of transformation, but consider for now what the effect would be of taking $\log_{10}$ of each value in the above dataset.

| Sample A | 1.48, 1.69, 1.87, 1.60, 1.80, 2.47, 1.78 | mean = 1.81 |
| Sample B | 1.73, 1.62, 1.98, 1.76, 1.81, 1.99, 1.94 | mean = 1.84 |

Now the means of both samples are fairly typical of the individual values in the samples, and if we compare these means we would conclude that typical distances flown between feeds were very similar in both suburbs. By transforming the values, we have reduced the influence of the extreme value. A comparison of the *mean of the transformed values*, i.e. $\log_{10}$(distance flown), is actually giving us a better comparison of which birds typically fly further than a straight comparison of mean distances flown would have.

There are some important features to note here:

- We have not changed the rank order of any of the individual values. The largest values in the original dataset are still the largest in the transformed dataset.

- We have not removed any values from the dataset. Every measurement or observation in the original dataset has a transformed value in the new dataset. We have not made any personal choices about which values to include.

- The influence of the sixth value in the first sample is very much reduced. It still contributes to our calculation of the mean but this single value no longer dominates the result we get.

I have used the example of a dataset with one extreme value in it to illustrate the usefulness of transforming data. Transformation can also be helpful when there are just a number of unusually high or unusually low values rather than one extreme value. In terms of frequency distributions these would appear as in Figure 6.3(a) and (b), though in environmental and biological sciences we seldom encounter distributions like Figure 6.3(a). Several slightly unusual values have the same effect as one extreme value; they tend to pull the mean towards them. Again, transforming the data will help to reduce their influence when we come to compare 'typical' members of the population. The more unusual or extreme a value, the more its influence will be reduced by transformation.

Usually in the process of transforming the data to obtain a set of values where the sample means are 'typical' of the individual values, we end up with datasets that meet the requirements for the most commonly used statistical tests such as *t*-tests and ANOVA (i.e. Normal distributions and equal variance). We can therefore use these tests to compare the means of the transformed values in our *samples*, to decide whether there really are differences between 'typical' members of the *populations* they came from as well.

When should you transform data?

If you intended to use a statistical test that assumes the data have a Normal distribution and/or equal variance and you find your data do not meet these assumptions, you have two main possibilities.

Use a non-parametric test

In some cases there may be a non-parametric test that is directly equivalent to the one you intended to use, but does not have such stringent assumptions behind it (Chapter 13). The disadvantage of using a non-parametric test will be that it is a less powerful test. However, if a non-parametric test does give a significant result, it will be straightforward to interpret, in which case little would be gained by transforming the data and using a parametric test instead.

Transform the data

Probably the most popular statistical methods for experiments and surveys in biological and environmental sciences are two-way or three-way ANOVA and regression. For these there are no straightforward equivalent non-parametric tests, so transforming the data may be the only possibility. However, transforming the dataset may make it harder to interpret. If we compared the transformed data in samples A and B above using a *t*-test, our conclusion would be how the *mean log distance flown* compared between the two suburbs not how the *mean distance flown* compared. However, whether we compare distances flown or logs of distances flown, a significant result could be interpreted as showing that the behaviour of the birds was not identical in the two suburbs.

How to transform data

If we have established that we need to transform our data in order to use the planned statistical test, we need to choose a suitable type of transformation. Different types of transformation are used in different situations. Some guidelines are given below, but looking at what others have done is very useful. You might find methods published in journals where similar situations were being studied which mention using a certain type of transformation. There is no guarantee that this was the correct thing to do, but it is a good clue.

To transform the data, start with the measured values (*y*) and perform the same calculation on each. We get a new set of values (*y'*). The analysis (*t*-test, ANOVA or regression) is then carried out on this new set of values. Some of the most commonly used transformations are described below.

Log transformation

This is probably the most commonly used. It can be used for data with a positively skewed distribution (Figure 6.3(b)), which often includes data where each value is a count (e.g. the number of insects on a leaf).

$$y' = \log_{10}(y + 1)$$

Natural logarithms, $\ln(y + 1)$, can be used instead of $\log_{10}$. One is added to all the values before taking the log to avoid the fact that we cannot take the log of zero. Since the test is to decide whether there is any *difference* between the populations, adding one to *every* measurement will not affect the outcome, or which sample has the larger values.

Square root transformation

The square root transformation is

$$y' = \sqrt{y}$$

It has a very similar effect to log transformation. For theoretical reasons, square root transformation is preferred when transforming data that come from a *Poisson distribution*. Poisson distributions occur when the data are counts of something resulting from a random process, such as things which arrive at random times or things which are spread randomly over an area.

Examples include datasets where (a) individual values are the numbers of daisy plants found in a series of randomly placed quadrat frames (if the daisies are distributed randomly in the study area), (b) individual values are the numbers of flies passing a detector in a series of 1 minute time intervals (if the flies arrive at random times).

If you can identify that you are studying a random process similar to (a) or (b), it is better to use a square root transformation. On a graph, a Poisson distribution with a mean value less than 5 looks similar to Figure 6.3(b) but becomes more symmetrical if the mean is higher. If the means of your samples are all greater than 5, the Poisson distribution is very similar to the Normal distribution, so no transformation is necessary. Remember, though, that if you have to transform one sample, you must transform them all.

Angular or arcsine transformation

The angular or arcsine transformation is often used when each value in your dataset is a *proportion* or *percentage* of something. It only applies where the minimum and maximum values are 0 and 1 (or 0 and 100%), e.g. the proportion of a pond covered with ice must be between 0 and 100%; however, the time taken to go by car as a proportion of the time taken by bus can be greater than 100%. If the original values (y) are expressed as percentages, they must be converted to values between 0 and 1 before transforming them (e.g. 72% = 0.72, 5% = 0.05).

$$y' = \sin^{-1}\sqrt{y} \quad \text{or} \quad y' = \arcsin\sqrt{y}$$

Examples include datasets where (a) individual values are the proportion of a leaf's surface which is necrotic (the dataset is then made up of values from a series of randomly selected leaves), (b) individual values are the proportion of seeds in a Petri dish which germinated (the dataset is then made up of values from a series of Petri dishes, each containing a number of seeds). In this latter case an alternative approach is to use a chi-square test to analyse the data (Chapter 12).

Box–Cox transformation

Box and Cox developed the following transformation:

$$y' = (y^\lambda - 1)/\lambda \qquad \text{(if } \lambda \neq 0)$$
$$y' = \ln y \qquad \text{(if } \lambda = 0)$$

It looks unfriendly but it has the advantage that it is possible to calculate the value of λ which best achieves a Normal distribution. This is not straightforward to do by hand, but a function is available on some computer packages to calculate the best value of λ for you. Once you have the value of λ it is fairly straightforward to put it in the above formula to calculate the transformed values, y'. This transformation achieves an approximately Normal distribution in a wide range of situations.

Once you have generated the transformed values, you will use this new set of values in your statistical test, so it is important to check that these new values meet the assumptions of the test. This can be done by graphing or otherwise examining the transformed values in the same way as for the original values (Section 6.2). Transformations do not always have the desired effect. If the transformed values do not meet the assumptions, go back to the original values and try another type of transformation. If none of the above transformations work, try other simple formulae such as $\sqrt[3]{y}$, $1/y$ or $1/y^2$.

   For data skewed to the left (Figure 6.3(a)), transformations such as y^2 or y^3 may be helpful. The aim is simply to achieve a dataset that meets the assumptions of the test by transforming every value in the same way. However, do not make it too complicated, because you have to interpret the result at the end. Note that if you use a transformation such as $1/y$ or $1/y^2$, transformed values will be in the reverse rank order to the original data; a smaller mean value of $1/y$ indicates larger values in the original sample.

   It is not *always* possible to achieve a Normal distribution using a simple transformation, in particular if a distribution has more than one peak in it – a *multimodal* distribution. This probably indicates that the population is actually made up of two or more different populations (e.g. male and female) which would be better treated as separate populations in the analysis.

PART II

STATISTICAL METHODS

Chapters 7 to 15 cover a range of statistical tests and methods and are intended as, essentially, a reference section. Each test or method is covered within one chapter and, except where cross-referenced in the text, it should not be necessary to refer to any of the other chapters in order to use it. Guidance on which test to use is given in Section 3.2 and in Appendix D. A basic understanding of the material covered in Chapters 1 to 6 will be helpful and is, to some extent, assumed.

7

t-tests and *F*-tests

7.1 Introduction

Both *t*-tests and *F*-tests are used in situations where we want to compare just two populations. These might be naturally occurring populations such as tree diameters in two different forests, or experimental populations such as the masses of caterpillars fed on two different diets. We can use a *t-test* to answer the question

How confident can I be that the two populations have different *means*?

Or an *F-test* to answer the question

How confident can I be that the two populations have different *variances*?

Another version of a *t*-test can be used to test whether the mean of one population is a particular value. This is called, appropriately enough, a one-sample *t*-test.

Remember that we will be studying only a small sample of individuals or points from each population but trying to decide whether the populations themselves differ. Statistical tests actually test how likely a particular statement is to be true, e.g. 'the populations have the same mean'. This statement is called the *null hypothesis* (Section 2.5) and the exact form of statement being tested varies from one type of test to another. The result of the test is a *P*-value, which tells us how likely the statement is to be true.

A major attraction of *t*-tests and *F*-tests is that they produce straightforward, easy-to-interpret results. This is an advantage not only for the person doing the work but also for the quality of the scientific information produced.

The tests as described here are all two-tailed tests, which is what you will usually need. However, it is possible to use them as one-tailed tests simply by replacing the alternative hypothesis with a one-sided alternative and halving the final *P*-value produced by the test (Section 2.5).

7.2   Limitations and assumptions

The following assumptions or requirements apply to *t*-tests and *F*-tests; they are explained in Chapter 6:

- Random sampling

- Independent measurements or observations

- Normal distributions

- Equal variance (*t*-tests only)

Both of the populations being compared should have Normal distributions. In a *t*-test the requirement for Normal distributions becomes less as the sample size increases. However, it is difficult to give a general rule as to what is acceptable, therefore it is safest always to ensure that the data have approximately Normal distributions. In a *paired t*-test it is the distribution of 'differences' which should be Normal (see below). In a one-sample *t*-test the distribution of the population being studied should be Normal. Data can often be transformed to bring about a Normal distribution if necessary (Chapter 6). The test would then be carried out on the transformed data. Alternatively an equivalent non-parametric test could be used (Chapter 13).

In a standard two-sample unpaired *t*-test, as explained in most basic statistics textbooks, the two populations being compared should have the same variance. (You could use an *F*-test on the samples to test whether they do.) However, with a modification of the calculations, it is possible to perform a *t*-test even when we cannot be confident that the populations have equal variances. This option is available on many statistical packages and can be used whenever we do not think it is safe to assume the variances are equal (even though they might be). Alternatively the data can be transformed to bring about equal variances (Chapter 6). There is no requirement for the populations to have equal variances to use an *F*-test – this is what we are testing for.

The calculations for both *t*-tests and *F*-tests can be carried out with as few as two values per sample, but the results from such a small sample size would be meaningless. We are trying to make conclusions about large and possibly very variable populations, and common sense should tell us that we cannot realistically judge what is a 'typical' member of a population or the spread of values in a population from just two measurements. A minimum sample size of 6 per population is therefore recommended.

   There is no strict requirement to have equal sample sizes for the two populations (except for the paired *t*-test) but there are theoretical reasons why the sample sizes should be similar. This can also be understood in commonsense terms. If you wanted to compare the mean weights of two populations of red deer, you would not choose to sample 4 from one population and 100 from the other. Knowing about one population very accurately is not much help if we have very little data about the group we are comparing them with. You would get a more reliable comparison by taking around 50 of each.

The data must be actual measurements or each value in the sample must be a count of something, so that we can calculate the mean for each sample. Data which are really scores on an arbitrary scoring system (1 = small, 2 = medium, etc.) are not suitable (Section 13.1).

7.3   *t*-tests

There are three types of *t*-test: (i) *unpaired t-test* to test whether two populations have the same mean; (ii) *paired t-test* to test whether two populations have the same mean, where each measurement in one sample is paired with a measurement in the other; and (iii) *one-sample t-test* to test whether the mean of a population is equal to some particular value.

Unpaired *t*-test

Null hypothesis

The null hypothesis (H_0) and alternative hypothesis (H_1) for an unpaired *t*-test are as follows:

- H_0: there is no difference between the means of the populations that the samples came from

- H_1: there is a difference between the means of the populations that the samples came from

Description of the test

An example of when we might use an unpaired *t*-test is:

To compare mean numbers of days spent in hibernation by hedgehogs in two areas of the country, using randomly selected samples of each. Suppose 10 hedgehogs in each area are captured, radio-tagged and released. Their movements are monitored throughout the winter to determine the number of days when they did not leave their nests.

The numbers of days spent entirely in the nest were as follows:

North	· 100	107	115	96	102	110	109	103	99	106	mean = 104.7
South	95	90	102	100	98	90	100	95	88	105	mean = 96.3

If you look back at Section 2.4, you will see that just by chance we can easily get samples with different means, even from the same, or identical populations (Figure 2.3(b)), so this might have happened here with the hedgehogs. And from Section 2.5 you will see that if we also calculate the likely range for the true population mean in each case, we can make a decision about how likely it is that the two samples did come from identical populations. If the difference between the sample means is large, and the 'margins of error' in these means are small, then we can be confident that the populations were different (Figure 2.4).

In Figure 2.4 we are left wondering whether the samples represented in (d) probably came from different populations or not. We can use a *t*-test to give us an objective way of deciding. The formula for a *t*-test simply gives us a number that reflects how big the difference between sample means is in relation to the 'margin of error' around the sample means.

The formula for an unpaired *t*-test can be summarized as

$$t = \frac{\text{(difference between sample means)}}{\sqrt{2 \times \text{(average variance of the samples)}/\text{(sample size)}}}$$

The value of *t* we get will be large if the difference between the sample means is large and the 'margin of error' around the sample means (represented by the bottom half of the equation) is small. The 'margin of error' around the sample means will be smallest if the populations themselves are not very variable, and if we have used a large sample size. If we get a large value for *t*, we can conclude that the samples probably did come from populations with different means.

How large is large enough? If we were to repeatedly draw pairs of samples at random from *identical* populations and calculate the *t*-value each time using the formula above, *t*-values greater than about 2 would occur less than 5% of the time because usually the two samples would have very similar means. So if our experimental data give us a *t*-value greater than about 2, we could conclude that there is < 5% chance that these two samples actually did come from identical populations.

If you refer back to Section 2.5, you will see that this would usually be considered strong enough evidence for us to conclude there was a significant difference. H_0 can be rejected and H_1 can be accepted. The exact value of *t* to use when deciding whether the populations differ, varies depending on the sample size. For a sample size as small as 6, the critical value of *t* rises to approximately 2.6. The correct values of *t* to use for each sample size are available in *t*-tables (found in the back of many statistics textbooks), or as functions on computers. If a *t*-test is carried out by computer, the *t*-value is effectively looked up for you, and the result of the test is given as a *P*-value directly, i.e. the probability of getting a result like this if the two samples actually came from populations with identical distributions.

The calculations for the hedgehog hibernation example are given in Appendix A.

Interpreting the results

The result of a *t*-test is a *t*-value, which can be looked up in tables or using a computer, to get an equivalent *P*-value. Here are some examples of how to interpret the results of an unpaired *t*-test:

- If the result of an unpaired *t*-test is a *P*-value > 0.05 and the sample means are males 26 m, females 19 m, we could conclude:

  There is no significant difference between the mean distances for males and females.

- If the result of an unpaired *t*-test is a *P*-value ≤ 0.05 and the sample means are males 26 m, females 19 m, we could conclude:

  The mean distance for males is significantly greater than for females.

For the hedgehog hibernation example we get $P = 0.004$, so we can conclude that there was a significant difference between the mean number of days spent in hibernation in the two areas. Looking at the sample means, we see that the mean was greater in the north, so we would conclude here that hedgehogs hibernated significantly longer in the northern area than in the southern area.

Paired *t*-test

Null hypothesis

The null hypothesis (H_0) and alternative hypothesis (H_1) for a paired *t*-test are the same as for the unpaired *t*-test.

Description of the test

An example of when we might use a paired *t*-test is:

To study whether SO_2 concentrations in the atmosphere are changing. We measure the concentrations of SO_2 at each of 12 randomly selected sampling sites in one year and then measure at the *same* locations in the following year. The first sample consists of the measurements made in the first year. The second sample consists of the measurements made in the second year. Each site contributes a *pair* of measurements, one in each sample.

The SO_2 concentrations in parts per billion (ppb) at the 12 locations were:

Site	1	2	3	4	5	6	7	8	9	10	11	12	Mean
Year 1	400	20	24	95	228	116	65	112	35	45	81	197	118.2
Year 2	345	8	29	81	204	140	36	75	47	5	65	187	101.8

If our dataset consists of a set of measurements such that each measurement in one sample is paired with a measurement in the other, we can use a paired *t*-test. The test begins by subtracting each measurement in one sample from its pair in the other. This gives us a set of *differences* which can be positive or negative.

Site	1	2	3	4	5	6	7	8	9	10	11	12	Mean
Difference	55	12	−5	14	24	−24	29	37	−12	40	16	10	16.3

If the two samples came from populations with the same mean, we would expect these differences on average to be zero. If, for instance, the values in the first sample came from a population with a higher mean, we would expect a lot of the pairwise differences to be positive. The question we need to ask is whether the *mean of the differences* that we find from our samples is far enough from zero for us to conclude that they probably came from populations with different means.

The formula for a paired *t*-test can be summarized as

$$t = \frac{\text{(mean of pairwise differences)}}{\sqrt{\text{(variance of the pairwise differences)}/\text{(sample size)}}}$$

As for the unpaired *t*-test, larger values of *t* mean that we can be more confident that two samples come from populations with different means. Again, how large *t* needs to be depends on the sample size and can be looked up in tables or using a function on a computer. Most computer packages do this as part of the *t*-test and give a *P*-value as the result of the test. This can be interpreted as for the unpaired *t*-test (see above).

In the SO_2 example there is a lot of variation between individual sampling locations, which would tend to mask the change over time that we are looking for. However, using a paired *t*-test we are comparing each individual in the sample only with itself at a different time. The test works only on the pairwise differences, not the measurements themselves; therefore, so long as the *effect of the treatment* (in this case *time*) on different individuals is fairly consistent, we will detect it. For this reason, in situations where we have paired measurements like this, a paired *t*-test is more powerful than an unpaired *t*-test and therefore a better test to use. The calculations for the SO_2 concentrations example are given in Appendix A.

Interpreting the results

The result of a *t*-test is a *t*-value, which can be looked up in tables or using a computer to get an equivalent *P*-value. Here are some examples of how to interpret the results of a paired *t*-test:

- If the result of a paired *t*-test is a *P*-value > 0.05 and the sample means are 114 points for test 1 and 116 points for test 2, we could conclude:

  There is no significant difference between the mean scores on the two tests.

- If the result of a paired *t*-test is a *P*-value ≤ 0.05 and the sample means are 114 points for test 1 and 116 points for test 2, we could conclude:

  The mean score on test 2 was significantly greater than on test 1.

For the SO_2 concentrations example we get $P = 0.030$, so we can conclude that there was a significant difference between the mean concentrations in the two years. Looking at the sample means, we see that the mean was greater in the first year, so we would conclude there had been a significant reduction in mean SO_2 concentration by the second year.

One-sample *t*-test

One-sample *t*-tests are not used as commonly as two-sample tests. With a one-sample *t*-test we can test whether the mean of a single population is equal to zero or to some other specified value.

Null hypothesis

The null hypothesis (H_0) and alternative hypothesis (H_1) for a one-sample *t*-test are as follows:

- H_0: there is no difference between the mean of the population and some fixed value.

- H$_1$: there is a difference between the mean of the population and a fixed value.

Description of the test

An example of when we might use a one-sample *t*-test is:

Ammonium sulphate, 3% enriched with ^{15}N is added as the only nitrogen source to barley plants growing in hydroponic culture. We want to check that 3% of the nitrogen taken up by the plants is also ^{15}N (i.e. that the plants do not discriminate in favour of or against ^{15}N). Ten barley plants are grown in individual containers, topped up with deionized water once per week. After 10 weeks, the percentage of ^{15}N as a proportion of total N in the remaining nutrient solution from each plant is determined.

The ^{15}N percentages measured in the individual containers of nutrient solution were:

3.01, 2.98, 2.97, 3.00, 3.03, 3.10, 2.94, 2.91, 2.99, 2.99   mean = 2.99

As explained in Section 2.4, we can use the data from our sample to calculate a range of values in which the population mean is most likely to lie; this is called the 95% confidence interval. If we calculate the 95% confidence interval for the above sample, we get 2.96–3.03. In other words, we believe these measurements probably came from a population with a mean somewhere in the range 2.96% to 3.03%. It is entirely plausible, therefore, that the population mean is 3.0% (our null hypothesis). This is how a one-sample *t*-test works. We *could* carry out this test by calculating the 95% confidence interval and seeing if our fixed value lies within it, but the following formula achieves the same thing:

$$t = \frac{(\text{sample mean}) - (\text{fixed value})}{\sqrt{(\text{variance of sample})/(\text{sample size})}}$$

The more the *sample* mean differs from the fixed value we are comparing it with, the larger the value of *t*. As for the other *t*-tests, the exact value of *t* at which we would start to conclude there was a difference for the *population* depends on the sample size. The *t*-value can be looked up in *t*-tables, although computer programs usually give an equivalent *P*-value directly.

The calculations for the ^{15}N enrichment example are given in Appendix A.

Interpreting the results

Here are some examples of how to interpret the results of a one-sample *t*-test:

• If we are testing whether a population has a mean of 1, and the result of a one-sample *t*-test is a *P*-value > 0.05 and the sample mean is 1.5, we could conclude:

The mean is not significantly different from 1.

• If we are testing whether a population has a mean of 1, and the result of a one-sample *t*-test is a *P*-value ≤0.05 and the sample mean is 1.5, we could conclude:

The mean is significantly greater than 1.

For the ^{15}N enrichment experiment we get $P = 0.63$, so we can conclude that we have no evidence that the plants discriminated in favour of or against ^{15}N.

Least significant difference

Least significant difference bears the delightful acronym LSD and so presents us with one of the few legal opportunities to use LSD. During a *t*-test, if we get a non-significant result, we cannot interpret this to mean that there is *no difference*, it is merely that we have not found a difference – think about this for a little while. We might not *find* a difference either because there isn't one, or because we didn't look very hard. Consequently, it would be helpful if we could back up a non-significant result with a measure of how hard we looked. The LSD is this measure. It can also be thought of as a measure of the sensitivity of our test.

  Suppose I ask two people to compare the heights of two trees (in fact the real heights are 10.66 m and 10.93 m). The first person tells me he found no difference, the second person tells me there is a difference. The reason turns out to be that one has measured to the nearest 1 m, the other has measured to the nearest 10 cm. In a sense they are both correct but one has used a more sensitive test. So it is when comparing populations using a *t*-test. If two people carry out the same experimental comparison, one may find a difference and the other not, because one is using a more sensitive test. The thing which makes a statistical test more sensitive is using a larger sample size.

  Unfortunately, it is difficult to predict the sensitivity of the test until you have already collected the results, because it depends on the variance of the populations, which you don't know until you have collected your data. Even so, the LSD is a useful figure to calculate after the test, particularly if you got a non-significant result, because it can help you to understand why you got this. The LSD is calculated from the values you have collected using a fairly simple formula (Box 7.1). If you find that the LSD was very small, your test was very

Box 7.1   LSD for an unpaired *t*-test

If we have carried out a *t*-test and found no significant difference, we cannot conclude that there was no difference between the means of the populations we were studying, only that if there was it was probably small. If it was *very* small, of course, then probably it was of no practical importance anyway. However, we can work out the smallest difference our test would have been likely to detect by calculating the least significant difference (LSD), and this will hopefully confirm that we are unlikely to have missed anything of practical importance.

The formula for an unpaired *t*-test was summarized in Section 7.3 as

$$t = \frac{\text{(difference between sample means)}}{\sqrt{2 \times \text{(average variance of the samples)/(sample size)}}}$$

This can be rearranged to give

$$\text{difference between sample means} = t \times \sqrt{2 \times \frac{\text{(average variance of the samples)}}{\text{(sample size)}}}$$

Usually in a *t*-test the difference between the sample means determines the *P*-value we get. If instead we choose a *P*-value, we can work backwards through the process to determine what difference between sample means this corresponds to.

We usually take $P = 0.05$, from which we look up the value of *t* with $2n - 2$ degrees of freedom, where *n* is the sample size. We would have already found the average variance of the samples while carrying out the *t*-test. Using this equation, therefore, we can work out what difference between sample means would give us a *P*-value of 0.05 in the test. Any difference between sample means greater than this would give us a *P*-value <0.05, so we can say this is the least difference which would be significant, or the least significant difference.

In general, sample means are close to but not exactly the same as the means of the populations they came from. Hence the LSD is also the smallest difference between *population* means that would be likely to give a significant result in the test.

This representation of the formula assumes equal sample sizes. Appendix A quotes a more general statement of the formula, which allows for unequal sample sizes.

sensitive and you could have expected to detect differences between the population means down to this value. If you find that the LSD was large, your test was not very sensitive, so there could still be quite large differences between the population means that your test simply didn't pick up. If the two sample means differ by greater than or equal to the LSD, the test will give a significant result. If the two sample means differ by less than the LSD, the test will give a non-significant result. Hence the LSD is, quite literally, the least difference that is significant.

Functions to calculate the LSD are not readily available on many statistical packages although the calculations are reasonably straightforward (Box 7.1). We usually calculate the LSD for a probability of $P = 0.05$ (i.e. the least difference between sample means which would give us a P-value ≤ 0.05). We could also calculate the LSD for other probabilities simply by using a t-value for the required probability in the same formula.

Here is an example of how to interpret an LSD.

- If the result of a t-test was a P-value > 0.05 and we calculated the LSD to be 6.4, we could conclude:

  Our test showed no *significant* difference between the means, and even if there really was a difference between the population means, it was probably less than 6.4.

The formula for calculating the LSD is given in Appendix A.

7.4   *F*-test

Null hypothesis

The null hypothesis (H_0) and alternative hypothesis (H_1) for an F-test are as follows:

- H_0: there is no difference between the variances of the populations that the samples came from

- H_1: there is a difference between the variances of the populations that the samples came from

Description of the test

An example of when we might use an F-test is:

In the hedgehog hibernation example for the unpaired *t*-test, we carried out a
t-test which assumed that the variances were equal. We could have, and should
have, checked initially that there was no reason to doubt this assumption, either
by graphing the data or by carrying out an *F*-test to compare the variances.

The values are as given on page 96. Section 2.4 describes how to calculate the
variance from a sample; here are the results:

- North: sample variance $= 33.34$

- South: sample variance $= 32.23$

When comparing means (e.g. using a *t*-test or ANOVA) we usually look at the
size of the *difference* between them. But when we are comparing variances, we
look at the *ratio* of the larger value divided by the smaller value; there are good
mathematical reasons for doing this. If the two variances are very similar, this
ratio will be close to one. If the ratio of sample variances is much larger than
one, we have evidence that the two population variances were different. How
large this ratio needs to be for us to conclude that the populations had different
variances depends on the size of samples we use. Consequently, the *F*-test takes
into account the ratio of one sample variance to the other, and the sample sizes
used.

The formula for an *F*-test is

$$F = \frac{\text{larger sample variance}}{\text{smaller sample variance}}$$

The *F*-value is simply the ratio of the larger variance divided by the smaller
variance. The critical values of *F* to conclude that the populations' variances do
differ can be looked up in *F*-tables or using a computer, taking account of both
sample sizes.

At the outset of our studies we do not usually know for sure which, if either, of
the populations will have the greater variance, so we want to use a two-tailed
test (i.e. at the outset we want a test of whether variance of A $>$ variance of B or
variance of B $>$ variance of A). If a statistical package is used, the result of the
test will usually be given directly as a two-tailed *P*-value, which is what we want.
If *F*-tables are used, the *P*-values obtained are one-tailed probabilities. Some
computer programs also give one-tailed probabilities as the result of an *F*-test,
in which case this should be stated in the program's output. To obtain the
two-tailed probability from a one-tailed probability, we simply need to multiply
the *P*-value by 2. See Section 2.5 for further information about one-tailed and
two-tailed tests. Appendix A gives the calculations for comparing the variances
in the hedgehog example.

Interpreting the results

Here are some examples of how to interpret the results of an F-test:

- If the result of an F-test is a P-value > 0.05 (i.e. the one-tailed probability is $P > 0.025$) and the sample variances are 264 and 189, we could conclude:

  There was no significant difference between the variances of the populations that these samples came from.

- If the result of an F-test is a P-value ≤ 0.05 (i.e. the one-tailed probability is $P \leq 0.025$) and the sample variances are 264 and 189, we could conclude:

  The first sample came from a population with a significantly greater variance than the second.

For the samples of hedgehog hibernation data we get $P = 0.961$, so we can conclude that we have no evidence that the populations had different variances, hence we are okay to use the t-test on these data.

7.5   Further reading

Besides their simplicity in use and in interpretation of the results, t-tests have the advantage that there are a number of possible non-parametric tests which can be used instead if the assumptions of the t-test turn out not to be met (Chapter 13). When the assumptions required for an F-test are not met by our data, Levene's test may be a useful alternative. In cases where we want to compare the means of more than two populations, ANOVA is the usual test to apply (Chapter 8). To compare more than two variances, Bartlett's test can be used if the samples come from Normal distributions or, again, Levene's test may be useful in other situations.

8

Analysis of Variance

8.1   Introduction

Analysis of variance (ANOVA) is used when we want to test whether there are differences between the means of several populations, based on samples taken from each population.

Some terminology

- *Variable*: the thing we measure, e.g. height, root length, potassium concentration

- *Factor*: the thing whose effects we are investigating, e.g. pH, age, colour, field

- *Levels of a factor*: the experimental treatments we use, e.g. 1, 2 and $3 \, mg \, l^{-1}$; red, green, pink and blue; field 1 and field 2

Two examples of how we use these terms are:

- If we are investigating the effect of soil pH on the height of wheat plants, with treatments of pH 4, 5 and 6, there is a single *factor* (soil pH) with three *levels* (4, 5 and 6) and the *variable* we are measuring is height.

- If we are investigating the effects of different liming treatments (2, 4, $6 \, t \, ha^{-1}$) and different soil types (clay, silty clay, loam, sandy loam) on soil pH, there are two *factors* (lime addition and soil type). Lime addition has three *levels* (2, 4, $6 \, t \, ha^{-1}$), and soil type has four *levels* (clay, silty clay, loam, sandy loam), and the *variable* we are measuring is soil pH.

One-way ANOVA

One-way ANOVA is like an extension of a t-test. In a t-test the null hypothesis we are testing is that two populations have the same mean. In one-way ANOVA the null hypothesis we are testing is that *all* the populations have the same mean. In fact, you *could* use one-way ANOVA to compare just two populations and the result would then be consistent with that of a t-test.

Multiway ANOVA

Two-way ANOVA is mathematically very similar to one-way ANOVA but its uses go much further. With two-way ANOVA we can test whether two different factors affect a particular variable, e.g.

Does age affect heart rate and does diet affect heart rate?

But we can also use it to test for an *interaction* between the two factors, e.g.

Is the effect of diet on heart rate different at different ages?

Interaction is described further in Section 8.4.

Three-way ANOVA, four-way ANOVA, and so on, also exist. We could therefore test whether age, diet and weight affect heart rate, and whether there is any interaction between these factors, and similarly if we had more than three factors. The procedure for carrying out the tests on a computer is usually the same, however many factors are involved. Many people are tempted to study a lot of factors in one experiment in the hope that something interesting will come out of all those numbers. However, the results often become dramatically more difficult to interpret as factors are added.

The ANOVA test itself will tell us if the factors interact in some way, but we would subsequently have to unravel this by comparing the mean values from each treatment combination. In an experiment with three levels of light, three levels of humidity, three levels of temperature and three levels of nutrients, one treatment combination would be: low light + medium humidity + low temperature + high nutrients. However, there are 81 such possible treatment combinations, which gives 3240 possible pairwise comparisons to carry out!

The results of one-way and two-way ANOVA are usually straightforward to interpret. The results of three-way ANOVA are usually possible to interpret but may be difficult, while the results of four-way ANOVA and higher are often very complicated to interpret. Consider breaking down studies involving more than three factors into several, more focused questions.

Repeated measures

Experiments and surveys often involve studying how things change over time. Datasets in which the *same* individuals or sampling locations are measured on a number of occasions are called *repeated measures*. This kind of study requires special considerations and is discussed in Chapter 11. However, if we want to study how populations change over time but we measure *different* samples of individuals on each occasion, these are not repeated measures in the statistical sense and we can treat *time* as one of the *factors* in ordinary ANOVA, as described in this chapter. Section 11.1 looks at this distinction in greater depth.

Univariate ANOVA vs. Multivariate ANOVA

This chapter describes techniques sometimes known as *univariate ANOVA*. That is to say, we may be studying the effects of several factors (e.g. altitude, steepness of slope, and underlying rock type), but we will only actually be measuring *one* variable (e.g. soil depth at each location). A related technique, *multivariate ANOVA*, can be used when we are measuring several variables in the study (e.g. Ca^{2+} concentration, Na^+ concentration, and soil pH at each location); this is described in Chapter 10. While it may sound appealing to include lots of factors and variables in an experiment, (because something has to affect something) this approach suffers from the fact that, ultimately, it is often difficult to unravel exactly what has affected what. Consequently, univariate ANOVA, as described in this chapter, is much more commonly used.

8.2   Limitations and assumptions

The following assumptions or requirements apply to ANOVA tests; they are explained in Chapter 6:

- Random sampling

- Independent measurements or observations

- Normal distributions

- Equal variances

All of the populations being compared should have Normal distributions. In general, whatever shape of distribution the groups do have, they will all be similar. Therefore we can make an overall assessment about Normality by looking at the distribution of the *residuals* from the test – the differences between individual measurements and the mean of the group they belong to (Box 8.1). To do this we carry out the ANOVA as planned, and in the process we generate and store a set of residuals; most computer packages will do this.

There will be one residual for each value in our dataset. We then check that these residuals have an approximately Normal distribution. If the residuals do have a Normal distribution we can use the results of the ANOVA test, otherwise we cannot. If the residuals do not have a Normal distribution, the data can often be transformed to bring about a Normal distribution (Chapter 6) and the test carried out again using the transformed data. Alternatively an equivalent non-parametric test can sometimes be used (Chapter 13).

The populations being compared should all have the same variance, i.e. they should all be equally variable. If this criterion is not met, data can be transformed to bring about equal variance (Chapter 6) or a non-parametric test can be used (Chapter 13).

The data must be actual measurements or each value in the sample must be a count of something, so that we can calculate the mean for each sample. Data that are really scores on an arbitrary scoring system (1 = small, 2 = medium, etc.) are not suitable (Section 13.1).

The *minimum* reasonable sample size depends on how many treatments there are. A good rule of thumb is to make sure you have at least *six* degrees of freedom in the error line of the ANOVA table; these terms are explained in the

---

Box 8.1   The meaning of residuals

If we have samples from several populations, we can calculate the mean of each sample.

Sample 1	4	7	9	12	mean =	8.00
Sample 2	21	20	15	18	mean =	18.50
Sample 3	114	110	110	117	mean =	112.75

The residuals are calculated by subtracting the mean of each sample from each of the individual measurements in that sample. So for sample 1 we get $(4 - 8)$, $(7 - 8)$, $(9 - 8)$ and $(12 - 8)$. Hence the residuals are:

Sample 1	−4.00	−1.00	1.00	4.00
Sample 2	2.50	1.50	−3.50	−0.50
Sample 3	1.25	−2.75	−2.75	4.25

These 12 values are the residuals we would obtain from an analysis of variance (ANOVA) on these samples.

next section. The error line in an ANOVA table may also be labelled 'residual' or 'within groups'. You can check how many degrees of freedom you will have before collecting your data by running the analysis with dummy results (Section 4.2). Increasing the sample size will increase the number of degrees of freedom in the error line. This is the minimum for the test to be meaningful. A larger sample size may be required to give you the precision you require (Section 4.2).

For one-way ANOVA the samples need not all be the same size (although to compare like with like it makes sense to try using similar sample sizes for all of the populations). For multiway ANOVA the calculations become more complicated if the samples are not all the same size, but provided you have a computer package which can deal with unequal sample sizes, this should not present a problem.

There is no theoretical limit to how many populations you can compare at once with one-way ANOVA and no theoretical limit to how many factors can be studied with multiway ANOVA. However, a significant result from these tests only tells you that the population means are probably not all the same. There are ways of deciding which populations differ from which others but these essentially entail comparing each one with each of the others; see pages 116 and 124. If you have included 100 populations in your study, you are in for an awful lot of work!

8.3 One-way ANOVA

Null hypothesis

The null hypothesis (H_0) and the alternative hypothesis (H_1) for one-way ANOVA are:

- H_0: the samples all come from populations with the same mean

- H_1: the samples do not all come from populations with the same mean

A significant result does not mean that all of the means differ from one another, just that they are not all the same.

Description of the test

An example of when we might use one-way ANOVA is:

Box 8.2   How ANOVA works

The idea behind ANOVA is really rather neat. If we have several samples taken from identical populations, we can estimate the variance of these populations in two different ways: (i) from the spread of values *within* the samples, and (ii) from how spread out the sample means are. Provided the samples *are* from identical populations, these two ways will give approximately the same result. If the samples come from dissimilar populations, they won't. The test involves estimating the variance in both ways and comparing these estimates.

Estimating the variance of the populations from the spread of values within the samples

One of the assumptions of ANOVA is that all the populations we are comparing have the same variance (Section 8.2); if they don't, we should not be using the test. Hence if we calculate the variances of each of our samples (Box 2.1) and take the mean of these variances, this will be quite a good overall estimate of the populations' variances. This is called the *within groups* estimate of variance; it may also be called the *residual* or *error* variance. For the millet example we get:

Variety 1	variance = 0.0149
Variety 2	variance = 0.0147
Variety 3	variance = 0.0112
Variety 4	variance = 0.0114
Variety 5	variance = 0.0108

The 'within groups' variance is 0.0126, the mean of these sample variances.

Estimating the variance of the population from the spread of the sample means

Box 2.2 explained that if we repeatedly take samples from a population, or from several identical populations, they would probably all give slightly different sample means, and the standard deviation of these means, called the standard error, is given by

$$\text{standard deviation of sample means, (or standard error)} = \frac{\text{standard deviation of individuals in the sample}}{\sqrt{\text{sample size}}}$$

We usually use this formula when we have only one sample, to give us a 'margin of error' for the mean. We are thus using the spread of individual values in the sample – the standard deviation – to estimate the spread of means we would expect if we were to take a number of *similar* random samples from that population – the standard error.

When we are using ANOVA we actually do have several *sample means*, so we can calculate their standard deviation directly, in the same way that we would calculate the standard deviation of the values in a sample (Box 2.1). We can therefore use a rearrangement of the above formula to estimate the standard deviation of the *individuals*:

$$\left(\begin{array}{c}\text{standard deviation} \\ \text{of individuals}\end{array}\right) = \left(\begin{array}{c}\text{standard deviation of sample} \\ \text{means (standard error)}\end{array}\right) \times \sqrt{\text{sample size}}$$

For the millet example:

Sample means	1.113, 1.140, 1.030, 1.393, 1.273
Standard deviation of sample means	0.1434
Sample size	3

Using the above rearranged version of our formula, we find

$$\text{standard deviation of the individuals} = 0.1434 \times \sqrt{3} = 0.2484$$

Variance, rather than standard deviation, is used in the test but variance is simply standard deviation squared. The 'between groups' variance is therefore $0.2484^2 = 0.0617$.

Comparing the variance estimates
Both the 'within groups' and 'between groups' variances are actually estimates of the variances of the populations our samples came from. If the null hypothesis is true, i.e. the populations all have the same mean, these estimates should tell the same story. If they don't, the null hypothesis is probably false.
To decide whether two variances are significantly different, we use an F-test (Section 7.4). In ANOVA we *always* divide the 'between groups' variance by the 'within groups' variance, regardless of which is larger.

$$F = \frac{\text{between groups variance}}{\text{within groups variance}} = \frac{0.0617}{0.0126} = 4.89$$

We would then look up this value up in F-tables or on a computer, using the one-tailed P-value, i.e. look up the P-value but don't double it as in an ordinary F-test. If the F-test gives a significant result, we can conclude that the two methods of estimating the variance did not give similar enough results, so the populations are probably not all the same.

A representative from a seed-producing company has been on a seed-collecting trip in the hope of finding improved varieties of millet. She wants to discover whether any of the local seed sources she has acquired produce superior yield to the others. If so, the better varieties will be investigated further. An experiment is set up with three replicated plots of each variety, and the yield per plot is determined at the end of the growing season. The order in which plots of the different varieties were positioned in the field was determined using a sequence of random numbers, i.e. this is a fully randomized design (Section 4.4).

The yields in tonnes per hectare ($t \, ha^{-1}$) were as follows:

Variety 1	1.14, 1.22, 0.98	mean = 1.113
Variety 2	1.01, 1.16, 1.25	mean = 1.140
Variety 3	0.95, 0.99, 1.15	mean = 1.030
Variety 4	1.51, 1.37, 1.30	mean = 1.393
Variety 5	1.19, 1.39, 1.24	mean = 1.273

Section 2.4 explained that if we take a series of randomly selected samples from identical populations, or even from the same population, each time we will probably get a different set of values, hence a different sample mean. Figure 8.1 shows three possible sets of results from an experiment comparing three populations, with four replicate measurements from each. In the situation shown in Figure 8.1(a) it seems quite feasible that the samples all came from identical populations. However, in Figures 8.1(b) and (c) it seems quite unlikely that *all* the samples are from identical populations.

What are we really looking at when we make this kind of judgement? We are actually considering how different the sample means are in relation to the amount of variation within the samples. This is how ANOVA works too. We can summarize this by saying that the samples are unlikely to have come from populations with the same mean if the variation *between* the sample means is large in relation to the variation *within* the samples.

ANOVA is simply an assessment of how likely it is that all of the samples come from populations with the same mean, considering both the variation between the means – the *between groups* or *treatments* variance – and the variation between measurements within the samples – the *within groups, residual* or *error* variance. A more detailed explanation is given in Box 8.2. The calculations are usually presented in a different way to those in Box 8.2 but they are entirely equivalent. In practice most people carry out the calculations on a computer but it is important to be able to interpret the output. The calculations for the millet yields example are given in Appendix A in the way that they are more usually set out.

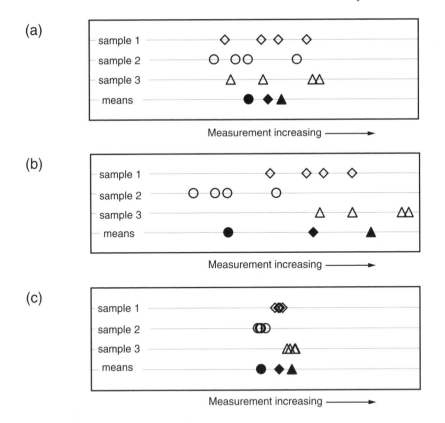

Figure 8.1   Dotplots of measurements made on some variable: End point represents one measurement (a) large variation within samples, small difference between sample means; (b) large variation within samples, large difference between sample means; (c) small variation within samples, small difference between sample means

Interpreting the results

The results of ANOVA are usually presented in an ANOVA table. The results for the millet varieties example are shown in the table below. You will see where the two variances calculated in Box 8.2 appear. To use ANOVA, it is not really necessary to understand all the figures in the table, but you might see them referred to in other texts so I have included a few explanatory notes. The main result of the test is the *P*-value in the final column.

ANOVA table

Source of variation	df	SS	MS	F	P-value
Between groups	4	0.2468	0.0617	4.8891	0.0191
Within groups	10	0.1262	0.0126		
Total	14	0.3730			

df = degrees of freedom, SS = sum of squares, MS = mean square
df for between groups = number of groups − 1
df for total = total number of measurements − 1
df for within groups = df for total − df for between groups
MS = (sum of squares)/(degrees of freedom)
F = (MS for between groups)/(MS for within groups)
MS are the estimates of variance referred to in Box 8.2
MS are what we compare in the test
$P = 0.0191$ is the P-value for $F = 4.89$ with 4 and 10 degrees of freedom
F-values have two associated degrees of freedom, df for between groups and df for within groups

Examples

Here are some examples of how to interpret a one-way ANOVA table:

- If the result of one-way ANOVA is a P-value > 0.05, we could conclude:

  There are no significant differences between any of the populations' means.

- If the result of one-way ANOVA is a P-value ≤ 0.05, we could conclude:

  At least two of the populations have significantly different means.

We would then need to decide which means differ, as explained below.

For the example of the millet varieties we obtained $P = 0.019$ from the ANOVA table. We can therefore conclude that the varieties probably do not *all* produce the same mean yield and we should investigate further to discover which were better than others.

Comparison of means after one-way ANOVA

If the result of the ANOVA is that there were no significant differences between the means, that is the end of the analysis. If the result is significant, we can conclude that at least two of the population means were probably different. In order to decide *which* of the varieties gave different yields, we can calculate the least significant difference (LSD); this is similar to the LSD for a *t*-test (Section 7.3). The calculations are given in Box 8.3. For the millet varieties we have LSD = 0.204.

If, even before seeing the results, we were particularly interested in whether two of the varieties differed (e.g. those with the most northerly and most southerly origins), we could calculate the difference between their sample means and compare this with the LSD. If the sample means differ by more than the LSD, i.e. if they differ by more than 0.204, we can consider these two varieties to be significantly different.

More commonly, we are not interested a priori in whether *particular* pairs of means differ; we are interested in *which* means differ. Our interest may focus on particular means after seeing the results, but this is different from having a reason to compare particular means *before* seeing the results. In this, more usual, case we need to apply a correction for multiple comparisons as described below.

Multiple comparisons

By convention we accept that two populations are probably different if a statistical test gives a *P*-value ≤ 0.05. This means that we accept up to a 5%

Box 8.3   Calculating the LSD for ANOVA

To calculate the LSD we need two values from the ANOVA table. For the millet varieties example in Section 8.3 we have

$$\text{Mean square for within groups (MSE)} = 0.0126$$
$$\text{Degrees of freedom for within groups (df)} = 10$$

The row 'within groups' may also be labelled 'error' or 'residual'. MSE stands for Mean Square Error.

We then need to look up the *t*-value for df = 10 and *P* = 0.05, from tables or using a computer; 10 was the degrees of freedom for 'within groups'; 0.05 is for 5% two-tailed probability. We obtain

$$t_{10,0.05} = 2.228$$

Finally we need the sample size, *n*, and here *n* = 3. The LSD at the 5% significance level for comparing treatment means can then be calculated:

$$\text{LSD} = \sqrt{2} \times \sqrt{\frac{\text{MSE}}{n}} \times t_{df,0.05} = \sqrt{2} \times \sqrt{\frac{0.0126}{3}} \times 2.228 = 0.204$$

chance of being wrong. If we carry out a series of tests and accept a 5% chance of being wrong in each one, there is much greater than a 5% chance that we will be wrong in at least one case. When we are trying to establish which means differ from each other, we are implicitly comparing each variety with every other one. With five varieties there are 10 possible pairwise comparisons, so if we accept a 5% chance of being wrong each time, the chances that we will be wrong in at least one case are quite high.

The ANOVA test itself is corrected for this, but when we come to make individual pairwise comparisons ourselves, we need to make a correction for the increased chances of being wrong in at least one case when we are making several comparisons. Statistical packages may offer a number of options for doing this, including the methods of Tukey, Duncan and Dunn–Šidák. Each of these applies slightly differing logic to how the correction should best be carried out, but for practical purposes any of these can be used. Dunnett's method can be used when one of the groups is a control and we want to compare each of the other groups with it, but not with each other.

If none of the above functions are available to you, the simplest correction for multiple comparisons to calculate by hand is the Bonferroni method. In this we start by deciding how many pairwise comparisons we are making, which can be found from

$$\text{number of possible pairwise comparisons} = \tfrac{1}{2}N(N - 1)$$

where N is the number of groups we are comparing. For the millet varieties we have $\tfrac{1}{2} \times 5 \times (5 - 1) = 10$ possible comparisons. We then divide the probability level that we would usually use to calculate the LSD (i.e. 5% or 0.05) by the number of possible comparisons. In this example our new probability to use to calculate the LSD is therefore $5\% \div 10 = 0.5\%$ (or 0.005). Finally we calculate the LSD, as described in Box 8.3 but using our new P-value, 0.005, rather than 0.05 when we look up the t-value to put in the formula (i.e. we use $t_{10,0.005} = 3.581$, rather than $t_{10,0.05} = 2.228$). This gives us an LSD of 0.328 (corrected for multiple comparisons).

Looking back at the sample means for the different varieties, in order of size we had:

Variety	3	1	2	5	4
Mean (t ha^{-1})	1.030	1.113	1.140	1.273	1.393

Only the first and last differ from each other by more than 0.328 ($1.393 - 1.030 = 0.363$), so we can say that in this case varieties 3 and 4 were significantly different (at $P \leq 0.05$) but none of the other varieties were significantly different from each other. Note that we are still testing for means which differ at $P \leq 0.05$ but we have corrected the LSD to allow us to make multiple comparisons at this probability.

Whether or not we need to apply a correction for multiple comparisons can be difficult to decide. Here are some pointers that indicate you *do* need to apply a correction:

- You want to know about *any* pairs of means in the experiment that differ from each other.

- You are interested in testing whether two means are different because from *looking at the results* it appears that they might be.

8.4 Multiway ANOVA

Multiway ANOVA is a very extensive subject. It is theoretically possible to use multiway ANOVA to study the effects of a large number of factors at once (e.g. pH, light, temperature, humidity, time of year, depth of water). However, it can be very difficult to interpret the results of multifactor studies if we want to know not only *which* factors have an effect, but also *what* their effect is. For example, we might want to know not only whether a pollutant is toxic to invertebrates but at what concentration it starts to become toxic. In such cases we will usually get clearer results by limiting the study to no more than three factors.

The most common use of multiway ANOVA in environmental or biological science is to determine whether two or three factors *interact*; that is, whether they act *together* to produce a greater or lesser effect than we would expect from their effects individually. For example, we might ask these questions: How does alcohol affect reaction times? And how does tiredness affect reaction times? These are questions about *main effects*, as they are called in statistical terminology. Main effects are the effects of individual factors, here alcohol and tiredness. We might also ask the question, Does alcohol affect reaction times more or less if you are tired than if you are wide awake? This question is about whether a person's reaction to alcohol is *affected by* tiredness. In statistical language, it is about a *two-way interaction*.

We can use two-way ANOVA to answer this kind of question. We could also test whether there was an interaction with a third factor, e.g. time of day, in which case the question becomes, Is the way alcohol and tiredness interact itself affected by time of day? This question is about a *three-way interaction*. You can see that if only two factors are involved, the question is fairly understandable, but when three factors are involved, even the question, let alone the result, is quite tricky to understand. If four or more factors are involved, things get correspondingly more complex again. Studying interactions of more than three factors is therefore only for the statistically confident or the foolhardy.

Factorial designs

When designing an experiment or survey to test for interactions, it is advisable to use a *factorial design*, or factorial set of treatments. This means that we will include measurements from every possible combination of the treatments or groups we are interested in. Suppose we have decided that a survey of aphids will include three different sites in a nature reserve (S1, S2 and S3), four different species of tree (T1, T2, T3 and T4) and insects collected at two different heights above the ground (H1 and H2). For a factorial design we need to ensure we include measurements from all the possible combinations:

S1, T1, H1	S1, T1, H2	S1, T2, H1	S1, T2, H2	S1, T3, H1	S1, T3, H2
S1, T4, H1	S1, T4, H2	S2, T1, H1	S2, T1, H2	S2, T2, H1	S2, T2, H2
S2, T3, H1	S2, T3, H2	S2, T4, H1	S2, T4, H2	S3, T1, H1	S3, T1, H2
S3, T2, H1	S3, T2, H2	S3, T3, H1	S3, T3, H2	S3, T4, H1	S3, T4, H2

If we had decided to use a sample size of 5, for example, we would need to sample five individuals at random from each of these treatment combinations. Note that a design can be both a factorial design *and* a fully randomized or randomized complete block design at the same time. The term 'factorial' refers to *which* combinations of treatments or groups are included in the study. The term 'fully randomized' or 'randomized complete blocks' refers to how these combinations are then arranged in time or space. For example, to actually carry out the above study we might collect one individual from each of the above treatment combinations in a random order on each day for 5 days. This would then be both a factorial design (because it includes all of these combinations) and a randomized complete blocks design, in which each day is a block (Section 4.4).

Null hypothesis

For multiway ANOVA we have several null hypotheses (H_0) and alternative hypotheses (H_1), all of which will be tested together. For two-way ANOVA we usually test three null hypotheses; for three-way ANOVA can have up to seven null hypotheses. The complexity quickly increases by including extra factors. A significant result for any of these hypotheses usually means that we have to carry out further comparisons to decide *how* the factors have an effect or interact. With four factors, a full interpretation of the results becomes even more complex still; in fact, the test would include up to 15 null hypotheses.

For two-way ANOVA

- H_0: factor 1 has no effect on the population mean
- H_1: factor 1 does have an effect on the population mean

- H_0: factor 2 has no effect on the population mean
- H_1: factor 2 does have an effect on the population mean

- H_0: there is no interaction between factor 1 and factor 2
- H_1: there is an interaction between factor 1 and factor 2

For three-way ANOVA

- H_0: factor 1 has no effect on the population mean
- H_1: factor 1 does have an effect on the population mean

- H_0: factor 2 has no effect on the population mean
- H_1: factor 2 does have an effect on the population mean

- H_0: factor 3 has no effect on the population mean
- H_1: factor 3 does have an effect on the population mean

- H_0: there is no interaction between factor 1 and factor 2
- H_1: there is an interaction between factor 1 and factor 2

- H_0: there is no interaction between factor 1 and factor 3
- H_1: there is an interaction between factor 1 and factor 3

- H_0: there is no interaction between factor 2 and factor 3
- H_1: there is an interaction between factor 2 and factor 3

- H_0: there is no interaction between factors 1, 2 and 3
- H_1: there is an interaction between factors 1, 2 and 3

Description of the test

An example of when we might use multiway ANOVA is:

The authorities governing a group of islands in the Pacific want to start a campaign to tidy up their beaches and need information on where to target their resources. It has been suggested that there is more rubbish on sandy beaches than shingly beaches. Another possibility is that people just notice it more on sandy beaches. An investigator wants to test this theory and also to compare the four main islands: Tinkewinke, Larlar, Dipsae and Sun Island. The survey consists of randomly selecting three beaches of each type on each island, and collecting the rubbish from a $20\,m \times 50\,m$ transect on each.

Some points to note are:

- There are two *factors*: island and beach type

- The factor island has four *levels*: Tinkewinke, Larlar, Dipsae and Sun Island

- The factor beach type has two *levels*: sandy and shingly

- This is a *factorial design*: *all eight possible combinations* of island and beach type are represented

- There are three *replicates*: three measurements from *each* combination of island and beach type are included

The masses of rubbish collected in kilograms (kg) were:

Beach type	Tinkewinke	Larlar	Dipsae	Sun Island
Sandy	1.5, 1.7, 1.0	3.6, 3.9, 3.1	2.6, 3.2, 2.1	2.0, 1.5, 1.9
Shingly	1.9, 1.2, 1.5	2.1, 2.5, 2.2	1.9, 2.3, 2.5	2.6, 2.1, 2.0

As for one-way ANOVA, the test makes an assessment of how likely it is that samples come from populations with the same mean by considering the variation *between* sample means relative to the variation *within* the samples. We can use two-way ANOVA to test for evidence of (i) whether the densities of litter differ overall on different islands, and (ii) whether the densities of litter differ overall for the different beach types. These are the *main effects* we will be testing in the ANOVA.

We might also ask, Is the difference between beach types the same on all four islands? Or are the differences between islands the same for sandy beaches as for shingly beaches? In fact, these are two ways of asking the same thing. In statistical terms, we are asking, Is there an *interaction*? The results are usually presented in an ANOVA table (see below). The calculations are given in Appendix A.

Interpreting the results

ANOVA table

Source of variation	df	SS	MS	F	P-value
Island	3	6.695	2.232	17.332	<0.001
Beach type	1	0.454	0.454	3.524	0.079
Island × beach type	3	2.501	0.834	6.476	0.004
Within groups	16	2.060	0.129		
Total	23	11.710			

The third row is the interaction between island and beach type
df for islands = number of islands − 1
df for beach type = number of beach types − 1
df for island × beach type = df for island × df for beach type
df for total = total number of measurements − 1
df for within groups = df for total − (df for island + df for beach type + df for interaction)
F = (MS for that factor or interaction)/(MS for within groups)
$P < 0.001$ is the P-value fo $F = 17.33$ with 3 and 16 degrees of freedom
$P = 0.079$ is the P-value for $F = 3.52$ with 1 and 16 degrees of freedom
$P = 0.004$ is the P-value for $F = 6.48$ with 3 and 16 degrees of freedom

Examples

Here are some examples of how to interpret a multiway ANOVA table:

- If the result is a P-value > 0.05, for any of the main effects we could conclude:

  Overall that factor had no significant effect.

- If the result is a P-value ≤ 0.05, for any of the main effects we could conclude:

  That factor had a significant effect.

- If the result is a P-value > 0.05, for an interaction we could conclude:

  There was no significant interaction between those factors.

If the result is a P-value ≤ 0.05, for an interaction we could conclude:

  There was a significant interaction between those factors.

For the example of rubbish on beaches, the ANOVA table leads us to the following conclusions:

- *Main effects*: for island, $P < 0.001$. We can therefore conclude that overall there was a significant difference between the densities of litter on at least two of the islands. For beach type, $P = 0.079$. We can therefore conclude that overall there was no significant difference between the densities of litter on the two types of beach.

- *Interaction*: for island × beach type, $P = 0.004$. We can therefore conclude that the difference between the densities of litter on different beach types was probably not the same on all of the islands, or the differences between the densities of litter on different islands was probably not the same for both beach types. These are two ways of looking at the same thing.

Comparison of means after multiway ANOVA

We can compare means after multiway ANOVA in the same way as described for one-way ANOVA (Box 8.3). In this case there are three possible LSDs, one to compare different islands, one to compare different types of beach (not particularly relevant here as there are only two), and one to compare particular island × beach type combinations. When calculating the LSD to use in each case, the only difference is in what we enter for the sample size.

The mean for each island is the mean of 6 measurements, the mean for each type of beach is the mean of 12 measurements, and the mean for each island × beach type combination is the mean of 3 measurements. Therefore, to calculate the LSD to compare islands we use a sample size (n) of 6 in the formula in Box 8.3, to calculate the LSD to compare beach types we use a sample size of 12, and to calculate the LSD to compare specific island × beach type combinations we use a sample size of 3.

In all cases the number of degrees of freedom to use when looking up the t-value is the number of degrees of freedom for the within (or residual or error) line of the ANOVA table (16 in this case). The LSDs we get are as follows:

- To compare islands, LSD = 0.439

- To compare beach types, LSD = 0.311

- To compare specific island × beach type combinations, LSD = 0.621

Multiple comparisons

The same argument applies to multiway ANOVA as to one-way ANOVA. If we are just interested in *which* means differ, or whether two means differ because the results look different, we are implicitly making all possible pairwise comparisons, so we should make a correction for multiple comparisons. Computer packages may offer a similar range of methods as for one-way ANOVA (see above). If these are not available, the simplest method is Bonferroni's method (see above).

For islands there are four means, so six possible pairwise comparisons. We should therefore use $P = 0.0083$ ($= 0.05/6$) rather than $P = 0.05$, when looking up the t-value to calculate the LSD. For beach types there are only two means and only one way of comparing them. The issue of multiple comparisons is irrelevant here. For specific island $\times$ beach type combinations there are eight means, so 28 possible comparisons, i.e. we should use $P = 0.0018$ ($= 0.05/28$) to calculate the LSD. The corrected LSDs are therefore:

- To compare islands, LSD $= 0.623$

- To compare specific island $\times$ beach type combinations, LSD $= 1.096$

As for one-way ANOVA, the decision whether or not to correct for multiple comparisons is not always easy. The safest choice if you are unsure is to make the correction. Means which appear different using the corrected LSD would also appear different using the uncorrected LSD.

Which means to compare

We can present the means of different groupings of individuals in the form of a table. Here is the table of means for litter collected in kilograms (kg).

Beach type	Tinkewinke	Larlar	Dipsae	Sun Island	Overall means
Sandy	1.40	3.53	2.63	1.80	2.34
Shingly	1.53	2.27	2.23	2.23	2.07
Overall means	1.47	2.90	2.43	2.02	

The effect of beach type was not significant, so we cannot conclude that one type or the other has more rubbish overall. We could use the corrected LSD calculated above (0.623) to compare the overall means for the different islands. Overall, beaches on Tinkewinke had significantly ($P \leq 0.05$) less rubbish than beaches on Dipsae or Larlar. Beaches on Sun Island also had significantly less rubbish than beaches on Larlar but not significantly less than beaches on Dipsae.

However, in this study it was found that there was a significant interaction between island and beach type. We can therefore conclude that the differences between islands were probably not the same for both beach types (or, if you prefer, the difference between beach types was not the same on all of the islands).

When the interaction is significant we should try to explain this by comparing the means for specific island × beach type combinations. In this case the correct LSD to use is 1.096, i.e. we can consider means for specific island × beach type combinations to be different from each other if they differ by 1.096 or more.

We find that on Larlar there was a significant difference between sandy beaches and shingly beaches, while on Tinkewinke, Dipsae and Sun Island there was no significant difference between the densities of rubbish on sandy and shingly beaches. We might also note that sandy beaches on Larlar had significantly higher densities of rubbish than any of the other types of beach except sandy beaches on Dipsae.

Analysing experiments with blocks

Section 4.4 explained that it was often a good idea to organize an experiment or survey as randomized complete blocks. Whether or not we take account of it in the analysis, it is useful to have randomized blocks because it ensures that unwanted variation will affect all of the populations we want to study to the same extent. However, if we can also identify *how much* this unwanted variation has affected the results, we can get a clearer picture of the effects of our experimental treatments themselves. We can achieve this by taking account of the blocks when carrying out an ANOVA.

There are various approaches to analysing experiments with blocks in the literature. The simplest is to consider the blocking factor (e.g. position in greenhouse) in the same way as one of the factors we are studying (e.g. fertilizer treatment, or watering treatment). Suppose we conducted an experiment with factors fertilizer treatment and watering treatment, and blocking factor position in greenhouse, we would analyse the results as a three-way ANOVA. The analysis must specify that you do not want to test for an interaction between the blocking factor and any of the factors of interest. Appendix C gives an example of how to do this using one popular statistical package.

It does not matter whether or not the blocking factor gives a significant *P*-value; we usually do not bother to report it. The point is that the *P*-values for the treatments we *are* interested in (in this case fertilizer and water) will then be calculated *allowing for* the effects of position in greenhouse, so we are more likely to get a significant result if differences really exist.

8.5   Further reading

Other approaches to ANOVA

The literature on ANOVA is vast. The technique can be extended to analyse studies with a variety of different blocking designs (Section 4.5), and to compare *groups* of treatments with other groups of treatments in the design, rather than just to compare individual means. Comparisons of groups of treatments are called *contrasts*. In practice the calculations for multiway ANOVA become less tidy if we have unequal replication or missing values, so an alternative method of carrying out the calculations called *general linear modelling* is often used in computer packages. For situations such as those described in this book, it gives identical results. An extension of this called *generalized linear modelling* can be used to carry out similar analyses in situations where the populations being studied do not have Normal distributions.

Analysis of covariance

Analysis of covariance (ANCOVA) can be used when we believe that some property, apart from the treatments themselves, might affect the variable we are measuring and we are able to make measurements of this property. For example, we are interested in comparing seed production of a plant species when it is growing among different vegetation types but we are not able to find equal numbers of plants in all of the vegetation types all at the same altitude. However, we suspect that altitude might itself have an effect.

One way to allow for this is to note the altitude where each plant was growing and include this as a *covariate* in the ANOVA. The analysis of covariance, as it is then called, will tell us whether altitude was linearly related to seed production, and if so, whether seed production differed between vegetation types after correcting the results for the effect of altitude. In effect, this test combines an ANOVA of the effects of different vegetation types on seed production, with a regression analysis (Chapter 9) of the effects of altitude on seed production.

9

Correlation and Regression

9.1 Introduction

Correlation and regression are used to study the relationship between two types of measurement made on the same individuals, e.g. mass vs. height. The term 'correlation' is usually used to refer to Pearson's product moment correlation. This gives a measure of how close the relationship between the two types of measurement, or *variables*, is to a *straight line*. Other types of correlation exist, including Spearman's rank correlation (Section 13.8), though these are usually referred to by their full names to distinguish them from Pearson's product moment correlation.

Regression is closely related to correlation and can be used to produce an equation to predict values of one of the types of measurement from the other. The term 'regression' is usually used to refer to simple linear regression, in which we test for a straight line relationship between the types of measurement. Other types of regression also exist but these are generally referred to by longer names to distinguish them, e.g. polynomial regression and nonlinear regression.

Correlation and regression should not be used without first examining the data using a scatter graph (Section 5.4). Data might in fact follow a curved relationship rather than a straight line relationship. The test itself would not reveal this, only that the data do not follow a straight line relationship. Perhaps even more concerningly, the test can suggest there is a weak straight line relationship, even though a graph would have shown that the relationship was really a very clear curve. The basic tests should not be used if there is clearly a relationship that is not a straight line. Options for dealing with curved relationships are discussed in Section 9.8.

Correlation and regression only test whether there is a relationship (or correlation) between two sets of measurements, not whether changes in one of the variables actually cause changes in the other. The number of doctors in a city is correlated with the annual number of deaths in a city. This does not mean that the doctors actually cause the deaths. It is just that bigger cities have more doctors and more deaths. Some other variable (in this case the size of the city) is the real cause of both the numbers of doctors and numbers of deaths. On discovering a correlation in a biological or environmental investigation, we should not therefore take this to mean that changes in one of the variables are necessarily causing changes in the other.

9.2   Limitations and assumptions

The following assumptions and requirements apply to correlation and regression; they are explained in Chapter 6:

- Random sampling

- Independent measurements or observations

- Normal distributions

- Equal variance (regression only)

For correlation the values for both types of measurement should have Normal distributions. For regression the *residuals* should have a Normal distribution (Figure 9.1). The easiest way to check this is to carry out the test and calculate the residuals; most computer packages have an option to save the residuals. Then check that the residuals have a Normal distribution. If they do, you can use the results of the test. If not, the data will need to be transformed and the analysis run again on the transformed data.

The equal variance assumption applies to regression analysis only. The variation of measurements above and below the fitted line should be the same

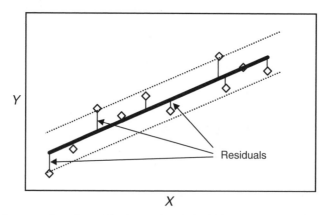

Figure 9.1   In regression, the residuals are the vertical distances between each of the points and the line of best fit. Points should be scattered fairly evenly above and below the line, and the amount of variation above and below the line should be fairly constant all along the line (indicated here by the parallel dotted lines)

all the way along the line (Figure 9.1). This is usually checked by looking at a scatter graph of the data.

The data for correlation and regression must be actual measurements or each value in the sample must be a count of something. Data that are really scores on an arbitrary scoring system (1 = small, 2 = medium, etc.) are not suitable (Section 13.1). Sample sizes less than 6 should be avoided because they cannot be relied upon to be representative of the population they came from.

9.3   Pearson's product moment correlation

In a correlation test we are studying the linear association between two variables. The two variables are interchangeable (i.e. the strength of correlation between X and Y is the same as the strength of correlation between Y and X).

Null hypothesis

The null hypothesis (H_0) and alternative hypothesis (H_1) for correlation are as follows:

- H_0: there is no straight line relationship on average between the variables
- H_1: there is a straight line relationship on average between the variables

Description of the test

An example of when we might use correlation is:

> Babies' sleeping patterns appear to be erratic. A study is carried out to discover whether the number of hours babies sleep by day is correlated with the number of hours they sleep the following night. A sample of babies is selected at random and for each one the number of hours slept during a day (8:00am to 8:00pm) and during the following night (8:00pm to 8:00am) is recorded.

The numbers of hours slept by each baby were as follows:

Baby	1	2	3	4	5	6	7	8	9
Daytime	6.3	4.2	10.3	7.1	8.4	9.5	5.0	3.5	6.4
Night-time	9.4	8.6	6.3	10.9	5.5	8.7	9.8	11.6	7.7

We should begin by plotting a scatter graph of the data (Figure 9.2).

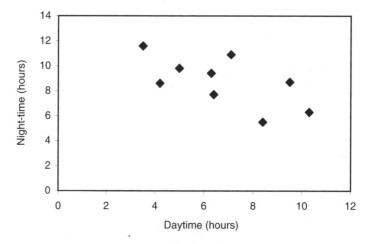

Figure 9.2   Scatter graph of hours slept by babies at night against hours slept the previous day

The graph suggests a possible straight line relationship. The sample size is rather small to show clearly whether the two sets of measurements have Normal distributions. However, from a visual inspection, the daytime measurements seem to be clustered more or less symmetrically around a mean of about 6.5 hours and the night-time measurements appear to be clustered more or less symmetrically around a mean of about 8.5 hours. Also it seems reasonable that most babies would sleep for a similar amount of time, with a few sleeping much longer and a few sleeping much less, so there is nothing here to make us doubt that they follow Normal distributions.

The test itself is simply a matter of applying a formula to the data to calculate the product moment correlation coefficient (r), which can then be looked up in tables, or by computer to give a P-value. The calculations for this example are given in Appendix A.

Interpreting the results

Correlation coefficient (r)

The correlation coefficient (r) is a measure of how well the points fit to a straight line. Values of r close to $+1$ or -1 indicate a close fit to a straight line, or a *strong correlation*. Values of r close to 0 indicate a very poor fit to a straight line, or *little* or *no correlation* (Figure 9.3).

There is no convention as to what values of r should be described as a strong or weak correlation. However, as a guide, the following terms could be used:

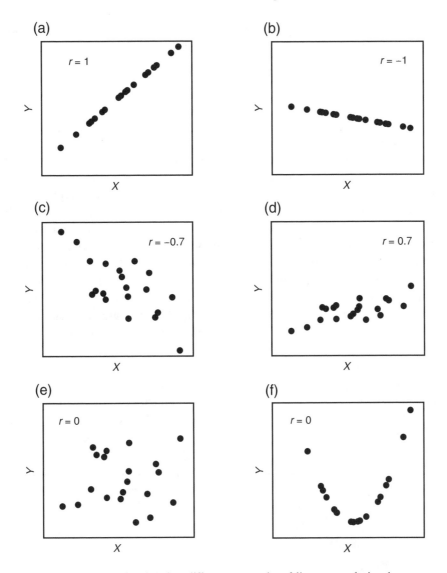

Figure 9.3 Scatter graphs showing different strengths of linear correlation between two variables, X and Y. The values of Pearson's product moment correlation coefficient (r) is shown in each case

r	$-1 \leftrightarrow$	$-0.9 \leftrightarrow$	$-0.6 \leftrightarrow$	$0 \leftrightarrow$	$0.6 \leftrightarrow$	$0.9 \leftrightarrow 1$
correlation		strong	weak	little or none	weak	strong

For the sleeping babies, the correlation coefficient is $r = -0.646$, a weak correlation in the sample data. The minus sign tells us that it is a *negative correlation*, i.e. values of night-time sleep tend to get larger as values of daytime sleep get smaller, and vice versa. This can also be seen by the fact that the line appears to slope downwards to the right in Figure 9.2.

P-*values*

The correlation coefficient, and indeed the graph, tells us whether or not the values in the *sample* show a straight line relationship on average. However, as with all the tests described in this book, our real interest is in the population the sample came from. We can never be certain that the sample is typical of the other members of the population, but we can say how confident we are. The *P*-value tells us how confident we can be that there is a straight line relationship between the two types of measurement on average in the *population*. It is therefore telling us something quite different from the correlation coefficient, which tells us how closely the points in the *sample* fitted to a straight line.The *P*-value is determined by the strength of the correlation (*r*) and by the sample size (Figure 9.4). It can be obtained by looking up the *r*-value in tables against the appropriate sample size, or directly from a computer. Frustratingly, some computer packages do not give the *P*-value, and the tables are not always readily available. An alternative is to use the regression command on a computer. Even though it may not be appropriate to use regression analysis per se on the data (Section 9.5), the *P*-value for the significance of the regression that you will obtain is the same as the *P*-value you would obtain by looking up the

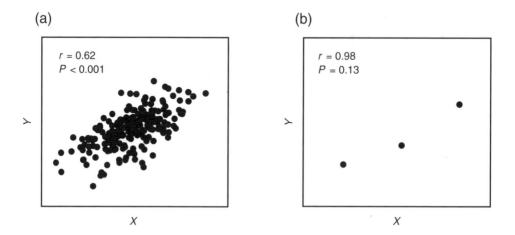

Figure 9.4 (a) Correlation with a large sample; the correlation is highly significant, even though *r* is quite low. (b) Correlation with a small sample; the correlation is not significant, even though the points in the sample lie very close to a straight line. The *P*-value tells us whether we can be confident there is a straight line relationship in the population, based on the evidence of the sample

value of *r* in tables. Ignore the rest of the output of the regression analysis as it is not relevant in a case like this.

Here are some examples of how to interpret the results from a correlation test:

- If the result of a correlation test is a *P*-value > 0.05, we could conclude:

  There was no significant correlation between the two types of measurement.

  We would conclude this even if the *r*-value seemed very close to +1 or −1.

- If the result of a correlation test is a *P*-value ≤0.05, we could conclude:

  There was a strong/weak significant positive/negative correlation between the two types of measurement.

  The value of *r* tells us whether it is a strong or weak correlation, and whether it is a positive or negative relationship (see above).

For the sleeping babies, we get $P = 0.060$. This is not significant, so we can say that there is insufficient evidence from this study to conclude that there is a correlation between the number of hours babies sleep by day and the number of hours slept the following night. In fact, judging from my two-month-old daughter, the length of time babies sleep at night isn't correlated with anything, no matter what you try!

9.4 Simple linear regression

Regression is similar to correlation in that we are testing for a linear relationship between two types of measurement made on the same individuals. However, regression goes further in that we can also produce an equation describing the line of best fit through the points on the graph.

When using regression analysis, unlike in correlation, the two variables have different roles. Regression is used when the value of one of the variables is considered to be dependent on the other, or at least reliably predicted from the other. You might guess that these would be called the dependent variable and independent variable respectively, and you would be right! The dependent variable might also be referred to as *Y* and the independent variable as *X*. When carrying out the analysis it is important to get the two variables the right way round. The regression of *Y* on *X* is not the same as the regression of *X* on *Y*.

In studies of correlation we select individuals entirely at random and *measure* both variables on each individual. Whereas in studies of regression we usually *choose* a set of fixed values for the independent variable (the one that we believe is controlling the other), e.g. we use temperatures of 5, 10, 15, 20 and 25°C then measure growth rates of a random sample of individuals at each temperature.

If it is not possible for us to control the independent variable we should try to ensure that the individuals in the sample give a fairly even coverage of the range of values in the population. Suppose I was studying how air quality on mountain summits is affected by altitude. I cannot choose what heights the mountains are. However, I can ensure that I include mountains spread fairly evenly across the range of heights in my study area by dividing them into smaller ranges (0–250, 251–500 m, etc.) and choosing, say, five mountains at random within each height range. This may be easier to understand if you imagine how it would appear on a scatter graph. For regression I want to get a fairly even spread of points all the way along the line, rather than a lot of points at low values of X (i.e. small mountains) and only a few at the high end of the X-axis (i.e. high mountains), even though there may be more small mountains in the area.

Any unusual or extreme values are likely to have a profound effect on the equation and the significance of the line fitted in the regression analysis; they are said to have high *leverage*. So far as possible, they should be avoided when setting up the experiment; it is better to use temperatures of 0, 8, 16, 24, 32 and 40°C than to use temperatures of 0, 1, 2, 3, 4, and 40°C. If you get any extreme values among the data you collect, consider carefully whether such points are genuinely part of the dataset or are due to some freak conditions or accident while collecting the data. If they are genuine you should include them, but it is probably useful to comment on them in the discussion of your results.

Section 9.5 looks in more detail at when to use correlation and when to use regression.

Null hypothesis

The null hypothesis (H_0) and the alternative hypothesis (H_1) for simple linear regression are:

- H_0: there is no straight line relationship on average between the variables

- H_1: there is a straight line relationship on average between the variables

Description of the test

An example of when we might use linear regression is:

To investigate the relationship between the length of time it takes eggs of a species of fly to hatch and the saturation deficit (i.e. humidity) of the air around them. A total of 210 eggs are divided at random into 10 similar clutches, 21 in each. Each of the clutches is kept at a different saturation deficit and all are maintained at 25°C. A record is made of the time it takes for the 11th egg in each saturation deficit treatment to hatch, i.e. the median hatching time. Section 13.1 gives further information on medians.

The results were:

Saturation deficit (kPa)	0	0.2	0.4	0.6	0.8	1.0	1.2	1.4	1.6	1.8
Median hatching time (h)	16.6	17.4	18.3	18.2	19.6	20.2	21.7	22.0	22.7	23.2

In regression analysis the independent variable, here the saturation deficit, should be controlled by the researcher, and we study the response of the dependent variable, here the median hatching time. The intervals for the values of the independent variable do not need to be exactly equal as here, but we should aim to have the values spread fairly evenly over the range we are interested in. In this example there is only one Y-value for each X-value – the median time for the eggs to hatch at that saturation deficit. In other studies we might have several Y-values for each X-value. Whether we have one Y-value for each X-value or several Y-values for each X-value, the regression analysis would be carried out in the same way.

The test involves calculating the equation of the line of best fit through the points (Figure 9.5). The position of the line is chosen to minimize the sum of the squares of the residuals (Figure 9.6). You could place the line by drawing a graph, moving the line around and measuring the vertical distances as shown until you found the place where the sum of squares of the residuals was least.

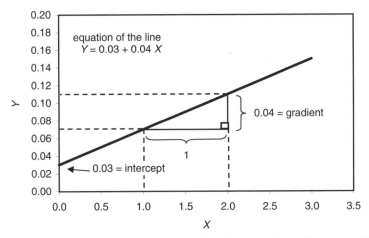

Figure 9.5   The meaning of the equation of a line. The equation tells us that if $X = 0$, $Y = 0.03$; and for any increase of 1 in X there will be an increase of 0.04 in Y

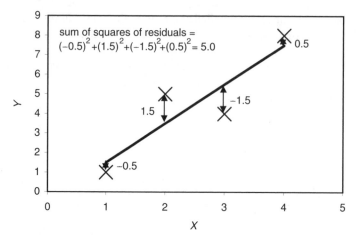

Figure 9.6 Calculating the sum of squares of the residuals in regression analysis

This would give you the same result (eventually). However, the position of the line of best fit can conveniently be found by using the regression formulae as given in Appendix A.

For the example of the flies' eggs, we start by graphing the data (Figure 9.7). There is no suggestion that this is a curved relationship, so we can proceed to use ordinary linear regression. Regression analysis produced the line shown, together with its equation. If you try entering any value of saturation deficit in the equation and calculating the median hatching time, you will see that it falls on the fitted line.

The regression formulae will always give us an equation, but simply obtain-

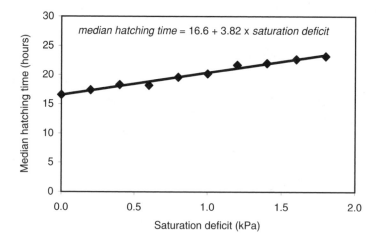

Figure 9.7   Median hatching times for flies' eggs at a range of saturation deficits. The line of best fit given by regression analysis is also shown

ing an equation does not in itself tell us whether it is really a useful description of what is going on in the population. The test therefore involves calculating a P-value to test whether we should conclude that there probably is a linear relationship in the population ($P \leq 0.05$), in which case the line is a useful description of it; or whether we shouldn't ($P > 0.05$), in which case the line is just an artefact of the calculations. The calculations for the hatching data are given in Appendix A.

Interpreting the results

The results of a regression analysis carried out by computer usually include the following information:

- The r^2 value is called the *coefficient of determination.*

- An 'ANOVA table' is usually included in computer outputs as part of a regression analysis. In this case the P-value tells us whether the regression is significant. i.e. whether there probably is a linear relationship between the X-values and the Y-values in the population.

- The *coefficients* give the equation of the line: $Y = $ intercept $ + $ slope $ \times X$. Each coefficient has a P-value which tells us whether we can be confident, from the sample, that that coefficient would be different from zero if we worked out the line of best fit for the whole population.

For the flies' eggs example the results look like this:

$r^2 = 0.982$

ANOVA table

	df	SS	MS	F	P-value
Regression	1	48.19	48.19	436.3	< 0.001
Residual	8	0.884	0.110		
Total	9	49.07			

	Coefficients	t-value	P-value
Intercept	16.6	84.74	< 0.001
Slope	3.82	20.89	< 0.001

Coefficient of determination (r^2)

Like the correlation coefficient (r) in correlation, r^2 tells us how well the points fit to a line, however r^2 values are always positive or zero. Values close to 1 indicate the points in the sample lay close to a straight line. Values near to zero indicate that the points were widely scattered around a line or showed no trend at all. As you might expect, r and r^2 are related: r^2 is what you would get if you worked out the correlation coefficient for the hatching data and squared it. However, they are used in different circumstances: r for correlation, r^2 for regression.

The coefficient r^2 is also equal to the proportion of variation in the dependent variable (median hatching times) that we could predict by knowing the values of the independent variable (saturation deficit) and the equation of the line. So r^2 is a useful indication of how good the fitted line is if we want to use it to estimate further values of Y. For the hatching data $r^2 = 0.982$, or 98.2% – r^2 is often given as a percentage. This indicates that the data fitted very closely to a straight line.

Significance of the regression

The *P*-value in the ANOVA table gives us the overall result of the test. Here are some examples of how to interpret the result of a regression analysis:

- If the result of regression analysis is a *P*-value > 0.05, we could conclude:

  There is no significant linear relationship between the dependent and independent variables.

- If the result of regression analysis is a *P*-value ≤ 0.05, we could conclude:

  There is a significant linear relationship between the dependent and independent variables.

For the hatching data, $P < 0.001$. We therefore have very strong evidence of a linear association between median hatching time and saturation deficit. Note that we actually only have evidence about how well the line fits *over the range of saturation deficits included in the study*. We should not assume that the relationship holds for greater saturation deficits as well.

Line of best fit

The coefficients give us the equation of the line of best fit, or the *least squares regression line.* For the hatching data, we have

$$\text{median hatching time} = 16.6 + 3.82 \times \text{saturation deficit}$$

The equation is only correct if we use the original units (i.e. hours and kPa). The regression line can be used to estimate further values of the dependent variable, median hatching time, for a given saturation deficit. The results also gave us a *P*-value for each of these coefficients. We can interpret them as follows:

- *Intercept* $= 16.6$ for this sample, $P < 0.001$ (highly significant). We can conclude that if we plotted this graph for the population as a whole, the intercept would not be zero. In other words, at zero saturation deficit the hatching time is not reduced to zero.

- *Slope* $= 3.82$ for this sample, $P < 0.001$ (highly significant). We can conclude that if we plotted this graph for the population as a whole, the slope would not be zero. In other words, there is a linear relationship between hatching time and saturation deficit. In simple linear regression, this *P*-value is always the same as the *P*-value for the overall significance of the regression.

Confidence limits

It is possible to give a *95% confidence interval* for the position of the *line* on the graph, similar to the 95% confidence interval for a population mean (Section 2.4). I do not go into the calculations here but some statistical packages will draw the limits on the graph as curved bands above and below the regression line. The true average relationship, if we could study the whole population, could be a straight line anywhere within these limits (with 95% confidence).

It is also possible to give a range of likely values for the median hatching time if we were to test a *further* cluster of the eggs at any particular saturation deficit. This is called the *95% prediction interval* and is a wider range than the 95% confidence interval for the line. This is because, having studied hatching times at a range of saturation deficits, we can be more confident about where the average relationship lies than where any particular future observation will lie. Some computer packages can also display the 95% prediction interval.

Estimating X-values from Y-values

Regression equations are ideal for estimating *Y*-values from *X*-values. However, another common use of regression is to fit lines to calibration data, such as a graph of peak height vs. concentration for a chromatograph. The height of the peak is the dependent variable, so using regression we will get an equation in the form

$$\text{peak height} = \text{intercept} + \text{slope} \times \text{concentration}$$

where *intercept* and *slope* are values given by the regression analysis. In this case we want to use the line to estimate values of concentration from peak height, so we must rearrange the equation to give

$$\text{concentration} = \frac{\text{peak height} - \text{intercept}}{\text{slope}}$$

Giving a confidence interval for the concentration in this case is unfortunately rather complicated, but is covered by Sokal and Rohlf (1995) for anyone who needs it.

9.5   Correlation or regression?

Both correlation and regression are tests of whether there is a straight line relationship between two variables. How do we decide which is the correct one to use? There are some cases where the answer is clear-cut:

- *Use correlation* if individuals are sampled at random from a population and we want to know whether the two variables, e.g. volume of stomach and diameter of mouth, are linearly related in the population on average.

- *Use regression* if one of the variables is controlled by the experimenter and we want to know whether the values of another variable respond linearly on average. For example, does number of species of bacteria respond linearly to amount of lime added to soil?

A lot of studies in biology and environmental science fall between these two, and controversies often arise about whether correlation or regression is appropriate. A problem often stems from the fact that we have a dataset which is most suitable for correlation analysis but we want to give an equation for the line of best fit. If both variables are measured on a random sample and neither is particularly the cause of changes in the other, it is difficult to say which should have the role of the dependent variable and which the independent variable. However, the line of best fit would be in a different place depending on which variable we treated as independent and which as dependent. Many texts recommend the use of model II regression in this case – the ordinary type of regression is called model I. Model II regression fits a line somewhere in between the lines we would get from regressing Y on X and X on Y, and might be described as the line which best describes the relationship in such cases. This is covered very readably, including the relevant calculations, by Fowler *et al.* (1998).

Model II regression has some disadvantages in practice. It is not available on all statistical packages, so calculating the equation of the line and testing the significance of the regression may mean getting involved in the calculations themselves. Also, if we actually want to use the fitted line to estimate values of one variable from the other, rather than just to characterize the relationship, it is said to be better to use ordinary model I regression, as described in Section 9.4.

Questions about our choice of method are most likely to arise if we simply measure the two variables on randomly selected individuals (ideal for correlation analysis) and then use regression analysis to fit a line to the resulting data. The best solution I can offer here is to think ahead about the type of results you will collect and try to carry out the study in such a way that it will be clear which technique you should use. We might, for instance, want to study the relationship between sulphate and chloride concentrations in a river. If we simply want to know whether there is a relationship, we could choose random sampling dates, measure the sulphate and chloride concentrations on each occasion, and calculate the correlation coefficient from these values.

However, if we want to determine an equation to predict values of sulphate concentration from chloride concentration, we should try to arrange things so that we have a fairly even spread of chloride concentrations over the range of values we expect to find in the river, and then measure the sulphate concentration in each case. We cannot control the chloride concentration, as we might in a laboratory experiment, but we could choose to sample at some times when we expect the chloride concentration to be low, some times when we expect it to be medium, and some times when we expect it to be high. This should give us a fairly even spread of chloride concentrations in our data.

Provided we can measure the chloride concentration fairly accurately on these occasions, the effect will be more or less the same as if we had chosen a set of values for the chloride concentration, and measured how the dependent variable (sulphate concentration) responded. We could therefore use regression analysis to analyse these data. It does not matter that the chloride concentrations used in the study are not spaced at exactly equal intervals along the X-axis, as long as they are quite well spread along the line.

9.6 Multiple linear regression

In simple linear regression we are testing for a linear association between one dependent variable and *one* independent variable. To test for a linear relationship between one dependent variable and *several* independent variables, we can use *multiple linear regression*. Extending the example of the flies' eggs, for instance, we might want to study how median hatching time (the dependent variable) depends on both saturation deficit *and* light intensity (two independent variables). For each clutch of eggs we would therefore have three values: saturation deficit, light intensity and median hatching time.

Both the saturation deficits and light intensities should be fixed by the experimenter and should be chosen so that they cover the whole range of interest. Unlike the factorial designs used in ANOVA (Section 8.4), it is not essential to have every saturation deficit × light intensity combination present in the design. Indeed, if we were measuring saturation deficit and light intensity on individuals in the natural environment, rather than controlling them in a laboratory, there might be no two individuals with exactly the same saturation deficit or light intensity. However, we should at least aim to include a good spread of combinations of saturation deficit and light intensity. For example, the analysis could give misleading results if we only included low saturation deficits at the low light intensities and high saturation deficits at the high light intensities.

If we repeated the experiment on the flies' eggs at two further light intensities and pooled all the results, we would get an output from the multiple regression analysis something like this:

$r^2 = 0.831$

ANOVA table

	df	SS	MS	F	P-value
Regression	2	128.19	64.35	66.34	< 0.001
Residual	27	26.19	0.970		
Total	29	154.88			

	Coefficients	t-value	P-value
Intercept	18.17	13.33	< 0.001
Slope (light intensity)	−0.0558	−1.27	0.216
Slope (saturation deficit)	3.58	11.45	< 0.001

This tells us that, overall, the regression is highly significant ($P < 0.001$), i.e. the line of best fit *is* a useful description of the average relationship between these variables. The equation of the line of best fit is

$$\text{median hatching time} = 18.17 - (0.0558 \times \text{light intensity}) + (3.58 \times \text{saturation deficit})$$

From the table of coefficients we can draw these conclusions:

- *Intercept* = 18.17 for this sample, $P < 0.001$ (highly significant). We can conclude that for the population as a whole the median hatching time would not be zero when the light intensity and saturation deficit were zero. This is not very useful information in this case but it might be useful in other situations.

- *Slope* (*light intensity*) = − 0.0558 for this sample, $P = 0.216$ (not significant). Although the equation suggests that median hatching time tended to decrease with increasing light intensity in the sample data (shown by the minus sign), we have insufficient evidence from this experiment to conclude that this is a real relationship. If we wanted to predict values of median hatching time, therefore, we might as well not bother measuring light intensity.

- *Slope* (*saturation deficit*) = 3.58 for this sample, $P < 0.001$ (highly significant). We can conclude that there is a significant positive relationship between median hatching time and saturation deficit (i.e. as saturation deficit increases, median hatching time increases).

It is theoretically possible to have any number of independent variables, but for the calculations to be carried out at all, the number of individuals in the sample must be at least two greater than the number of independent variables.

Adjusted r^2

If we were interested in what controls plant growth rates, we might measure the growth rate of 100 plants in a wide range of different locations. For each plant we could also measure a number of aspects of its environment (e.g. mean light intensity, mean temperature, soil depth, altitude). We could start by using simple linear regression to produce an equation to predict, say, growth rate from mean temperature. This would yield a value of r^2, the proportion of variation in growth rate that we can predict simply by knowing mean temperatures. We could then use multiple regression to produce an equation to predict, say, growth rate from mean temperature and soil depth. Again this would yield a value of r^2, now the proportion of variation in growth rate that we can predict knowing mean temperatures and soil depths.

However, it turns out that, by a mathematical quirk, every time we add a variable in this way, the value of r^2 increases, which might lead us to think that all the variables we try are useful for predictive purposes. The 'adjusted r^2 value' is, as the name suggests, the r^2 value adjusted to allow for this effect; it is given by most computer packages. We can tell whether adding additional independent variables is really useful by checking whether the adjusted r^2 value increases. Note this applies to cases where we are adding to the regression new variables measured on the *same* individuals. In our egg-hatching example above, additional batches of eggs were added into the experiment when light intensity was included as a factor in the study. In such cases it is possible for the r^2 value to go up or down.

9.7   Comparing two lines

So far we have looked at testing whether there is a straight line relationship between two or more variables in one population. We could of course do this separately for any number of populations. This section considers how to answer the question, Are there different straight line relationships between the two variables in two different populations?

Figure 9.8 shows the results of a survey of former university lecturers admitted to mental institutions, and the amount of lecturing they had been doing. We have samples from two populations, science lecturers and arts lecturers, and we want to test whether the relationship between age on admission to an institution and hours per week of teaching is different in the two populations. We can see that the relationship is different in the *samples*; the test is for whether there is also a difference between the *populations*.

Many computer packages do not have a convenient function to compare regression lines. The calculations can be done by hand but are not inviting. An alternative is to use multiple linear regression. This involves setting up the dataset in a particular way. The Y and X values are entered in adjacent

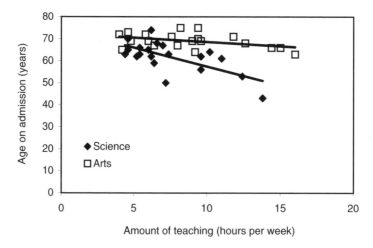

Figure 9.8   The relationship between age on admission and number of hours per week of teaching in their final year of work, for university lecturers from science and arts subjects admitted to psychiatric hospitals

columns, with the data for one population immediately below the data for the other. A further two columns are then added as shown below for the lecturers data:

(1) Age on admission	(2) Hours per week	(3) Set	(4) Hours per week × set	
68	6.6	0	0	
59	6.4	0	0	
56	9.6	0	0	data for science
⋮	⋮	⋮	⋮	lecturers
65	6.0	0	0	
62	9.6	0	0	
72	4.0	1	4.0	
75	8.2	1	8.2	
70	9.4	1	9.4	data for arts
⋮	⋮	⋮	⋮	lecturers
75	9.4	1	9.4	
66	14.4	1	14.4	

There are four points to note:

- *Age on admission* is the dependent variable here.

- *Hours per week of teaching* is the independent variable.

- In the third column enter a zero against all the values from one population (science lecturers), and a 1 against all the values from the other population (arts lecturers); it does not matter which population you put first.

- Multiply column 2 by column 3 and put the results in column 4.

Now carry out a multiple regression test using the age on admission as the dependent variable (the values on the Y-axis), and the other three columns of numbers as the independent variables. As for the example of multiple linear regression in Section 9.6, you will get a P-value for the intercept and a P-value for each of the independent variables.

	Coefficients	t-value	P-value
Intercept	74.61	25.17	< 0.001
Hours per week	− 1.72	− 4.59	< 0.001
Set	− 2.19	− 0.55	0.586
Hours per week × set	1.33	2.85	0.007

From this we can obtain the equations of the two lines:

Science lecturers (set 0)

$$age = 74.61 - 1.72 \times hours\ per\ week$$

Arts lecturers (set 1)

$$age = (74.61 - 2.19) + (-1.72 + 1.33) \times hours\ per\ week$$
$$= 72.42 - 0.39 \times hours\ per\ week$$

We can now ask the question, Are the lines different? But what we have to realize is that the lines can be different because they have different intercepts and/or because they have different slopes. The third line in our results table gives us the difference in intercepts, -2.19, which is not significant ($P = 0.586$). I have chosen to call this variable 'set', but it can be given any name.

The fact there is no significant difference in the intercepts tells us that we have no reason to believe the two groups of lecturers would have been admitted at different ages if they had done no lecturing. The fourth line, for hours per week × set, gives us the difference in slopes, 1.33, which is significant ($P = 0.007$). Combined with the evidence from the graph, we can therefore conclude that hours spent in the lecture theatre had a significantly greater effect on the science lecturers than the arts lecturers.

9.8   Fitting curves

If a scatter graph of the data shows that the relationship is clearly not a straight line, there is little point fitting a straight line to it. If we just want to test for a correlation between two variables, knowing that the relationship is not a straight line, Spearman's rank correlation is often suitable (Section 13.8). If we want to fit a line to data that show a curvilinear relationship there are several possible options.

Transforming axes

If the data appear as in Figure 9.9(a) or (b), using logarithms of Y (to any base) as the dependent variable will lead to a straight line relationship, which can be tested with simple linear regression (Figure 9.9(c) and (d)). In effect we would test for a linear relationship between log Y and X, rather than between Y and X.

Similarly, curves in other directions can be 'straightened' by using logarithms of X as the independent variable. If in doubt, try plotting Y vs. log X or even log Y vs. log X and seeing if this produces a straight line.

Other simple transformations can also be used if they achieve a straight line. For example, we could test for a linear relationship between Y and $\sqrt{X}$,

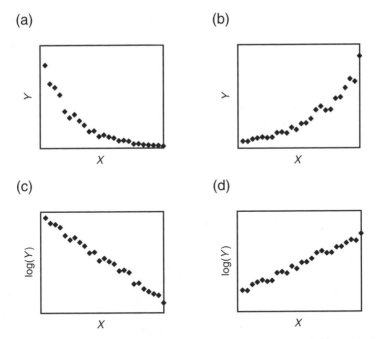

Figure 9.9 (a, b) Examples of curvilinear relationships between two variables, X and Y. (c, d) The effects of log transforming the Y-axis. (a and c) and (b and d) show the same data before and after transforming the Y-axis

between Y and $\sqrt[3]{X}$, between Y and X^2, between $1/Y$ and X, and so on. Our aim is just to find some simple relationship between the two variables, which might then be used to predict one from the other.

Polynomial regression

If the data appear as in Figure 9.10(a) or (b), or any part of these curves, use both X and X^2 as the independent variables. If you are using a computer package to carry out the analysis, you should have one column containing the X-values, then create an extra column containing the squares of each of these values, and then carry out a multiple linear regression using the columns of X and X^2 values as two independent variables (analogous to using light intensity and saturation deficit as two independent variables in the egg-hatching example).

In this case the significance of the regression tells you whether there is a quadratic relationship between Y and X, and the line of best fit will have the form

(a)   (b)

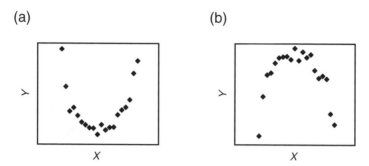

Figure 9.10   Examples of quadratic relationships between variables X and Y

$$Y = a + bX + cX^2$$

It is possible to extend this idea by adding a further column containing values of X^3, which are then included as a third independent variable. In this case the significance of the regression tells you whether there is a cubic relationship between Y and X. Further extensions are possible in the same way. By including X^4 and X^5 as well, it is possible to get lines that fit very closely to many shapes of curvilinear relationship. This can be useful if we need an equation to put in a mathematical model. However, it usually does little to explain the biological or physical mechanisms behind a relationship. Biological processes might react to aspects of their environment such as temperature or pH with a logarithmic or quadratic relationship, but cubic and higher-order polynomial relationships are usually the result of several underlying processes added together.

Nonlinear regression

All of the methods covered above are actually types of linear regression. When transforming the axes we subsequently test for a straight line relationship between the transformed variables, and in polynomial regression we are actually testing for straight line relationships between Y and X^2, between Y and X^3, etc.

It is also possible to test for curvilinear (i.e. curved) relationships between two variables, using nonlinear regression. If you are using a statistical package that will do this, you need to specify what sort of curvilinear relationship you want to test for. The analysis will then usually give you an equation for the line, the overall significance of the regression, and the significance of each of the coefficients for the fitted line, as for multiple linear regression.

Some types of curvilinear relationship commonly found in biological and environmental systems are shown in Figure 9.11 together with the forms of equations for the lines.

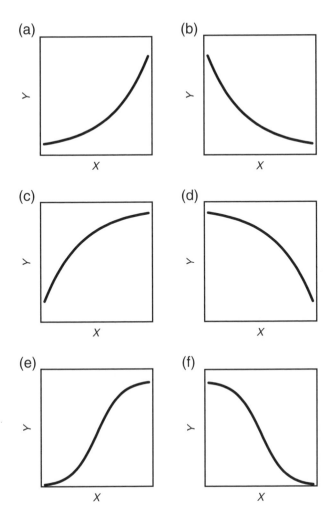

Figure 9.11   Types of curvilinear relationship commonly found in biological and environmental science: (a–d) exponential curves $Y = a + b\exp(cX)$ or $Y = a + bc^X$. (e, f) logistic
curves $Y = a + b/\{1 + \exp[-c(X - d)]\}$; a, b, c and d are constants (which may be positive
or negative), usually obtained from nonlinear regression analysis

9.9   Further reading

There are many extensions to the basic ideas behind regression analysis. Some
that may be useful in biological or environmental science are stepwise regression and best subsets regression. These can help us to choose which of the
independent variables we have measured, or which combinations of these
variables are actually useful to include in a regression equation.

Stepwise regression

Stepwise regression involves trying different variables in the regression systematically and retaining or removing them in relation to whether they really improve the power of the equation to predict changes in the dependent variable. The procedure requires the user to make some choices about when variables should be added or removed, so it requires more statistical knowledge than simple linear regression.

Best subsets regression

Best subsets regression identifies which would be the best of the independent variables to use if you were only prepared to use one variable to predict changes in the dependent variable, then goes on to identify the best pair to use if you were only prepared to use two, the best three to use if you were only prepared to use three, and so on. The procedure is simple to use on a computer, but take care not to accept the results blindly. The selection of variables is a purely mathematical process. The best combinations from a mathematical viewpoint may be a lot harder to measure in practice, or make less scientific sense than combinations you would choose with a little practical knowledge.

10

Multivariate ANOVA

10.1   Introduction

Chapters 7 and 8 describe how we can use t-tests and ANOVA to test whether a characteristic such as mean height differs between different populations or is affected by experimental treatments. Tests such as t-tests and ANOVA are said to be *unvariate methods* because they only consider one type of measurement at a time. If we were also interested in other characteristics (e.g. metabolic rate and mass), we would need to carry out similar tests on each of these *variables* separately. We would first test whether heights differed between the populations, then test whether metabolic rates differed, and then test whether masses differed.

An extension of the ideas behind t-tests and ANOVA is to test whether populations differ if we consider *several* variables at the same time. The method we use to test this is called *multivariate ANOVA*, or *MANOVA*. MANOVA is one of a class of methods that consider several types of measurement made on each individual together. These are called, not surprisingly, *multivariate methods*. Two further multivariate methods are principal component analysis (Chapter 14) and cluster analysis (Chapter 15).

MANOVA is less commonly used in environmental and biological sciences than t-tests or ANOVA because our investigations are usually aimed at identifying *which* aspects of populations are affected by their environments. MANOVA often gives too general an answer. Simply establishing that populations differ is often not very helpful; we would probably also want to know how they differ. However, MANOVA tests are sometimes useful as one part of the analysis of a multivariate dataset. We would often combine this test with graphical presentation of the data and/or analysis using another multivariate method such as principal component analysis or cluster analysis. We could also carry out ordinary ANOVA tests on each of the types of measurement individually.

Note that multivariate datasets do not necessarily consist of unlike types of measurement such as height, metabolic rate and mass. A multivariate dataset could also consist of measurements of the amount of light reflected from the same samples of leaves using light at 400 nm, at 410 nm, at 420 nm, etc. The amount reflected at 400 nm then constitutes one variable, amount reflected at 410 nm another variable, and so on. MANOVA might usefully be used in this case to test whether different populations of leaves have different overall reflection 'characteristics', rather than whether they reflect different amounts at any specific wavelength.

Similarly, a multivariate dataset could consist of a set of measurements made on the same samples on successive occasions. Just as we could measure the mass, length and thickness of a sample of leaves and consider them as three separate variables, we could measure the photosynthesis on day 1, the photosynthesis on day 2 and the photosynthesis on day 3 of a sample of leaves, and consider them as three separate variables. Datasets like this are called repeated measures. Using MANOVA to analyse this type of data is described in Chapter 11, along with other options for repeated measures data.

10.2   Limitations and assumptions

The following assumptions and requirements apply to MANOVA. These are similar to the common assumptions described in Chapter 6 but with some modifications to apply them to a multivariate dataset, as described below.

- Random sampling

- Independent measurements or observations

- Normal distributions

- Equal variances

MANOVA tests require random sampling and that the individuals or individual points included in our samples must be acting independently. However, on each individual we will measure several properties (e.g. its height, its metabolic rate, its mass). We cannot use a set of heights from one sample of individuals, and metabolic rates and masses from other samples of individuals, even if these come from the same populations.

Statistical texts usually state that MANOVA has an assumption that the measurements within each population have a *multivariate normal distribution* and equal *variance–covariance structures*. These are analogous to the assumptions of Normal distributions and equal variance for *t*-tests and ANOVA (Section 6.2). However, the texts are generally unhelpful about how to test for these conditions except to say that a number of (rather complex) tests do exist in the statistical literature.

If you are using a computer package that provides tests for these assumptions then it makes sense to use them. Otherwise a sensible practical approach that will provide an approximate test for these conditions is as follows.

Avoid including any variables, i.e. any types of measurement, that have a curvilinear relationship with any of the other variables (Section 9.8), e.g. leaf area is often proportional to leaf length squared. If you have variables such as these, either leave one or other out of the analysis, or use a measurement such as $\sqrt{\text{leaf area}}$ instead. The two variables, $\sqrt{\text{leaf area}}$ and leaf length, would then have a linear relationship, so both could be included.

Next consider each type of measurement, i.e. each variable, in turn and check that it meets the assumptions of ordinary ANOVA. For example, if your dataset consists of measurements of height, metabolic rate and mass, first check the height measurements have Normal distributions and constant variance, then check the metabolic rate measurements have Normal distributions and constant variance, and finally check the mass measurements have Normal distributions and constant variance, as if you were going to carry out ordinary ANOVA on each of them individually (Section 8.2). If any of the variables do not meet these criteria, transform all the measurements for that variable.

Different variables may require transforming in different ways or not at all. If you do have to transform any variables, you will need to recheck that none of them now has a curvilinear relationship with any of the other variables (see above). Once each of the variables individually meets the requirements for ordinary ANOVA, carry out the MANOVA.

While this approach does not guarantee that the dataset will exactly meet the requirements of MANOVA, texts generally consider MANOVA to be fairly unaffected by small departures from multivariate Normal distributions and constant variance–covariance structures.

The data must be actual measurements or each value in the sample must be a count of something, as the test is actually based on comparing means, as for ordinary ANOVA. Data that are really scores on an arbitrary scoring system (1 = small, 2 = medium, etc.) are not suitable (Section 13.1).

There is no theoretical limit to the number of variables that can be included in a multivariate ANOVA; we could include the length of every bone in an animal's body in one MANOVA test. However, the calculations involved become vast when a large number of variables are involved, and some computer packages may not handle them.

There can also be problems with not having enough degrees of freedom for the analysis to be carried out (Section 4.2). Essentially this occurs when our dataset consists of many different variables measured on a relatively small number of individuals. The simplest way to check that you will have enough measurements made on enough individuals is to carry out the analysis using dummy data before collecting the real data (Section 4.2).

10.3   Null hypothesis

The null hypothesis (H_0) and alternative hypothesis (H_1) for multivariate ANOVA are analogous to those for ordinary ANOVA (Chapter 8). When we are only testing for the effects of one factor we have these hypotheses:

- H_0: the samples all come from populations with the same mean vectors

- H_1: the samples do not all come from populations with the same mean vectors

Populations have the same mean vectors if the means for *all* of the variables included in the analysis are the same. MANOVA can also be used to test the effects of *several* factors and whether they interact, analogous to ordinary multiway ANOVA. We would then have several null hypotheses and alternative hypotheses as in Section 8.4.

10.4   Description of the test

An example of when we might want to use MANOVA is:

> Suppose we want to determine whether three isolated groups of Scots pine trees in neighbouring valleys are genetically distinct or not. In terms of any particular characteristic, such as height, density of branching or bark thickness, it may be impossible to distinguish between them, but considering all of these variables together, we could test whether the populations differ overall. This would tell us whether the populations have evolved separately over a long period or whether they were part of the same woodland until fairly recent times.

For clarity I consider a dataset with only four trees in each group, and in which only three variables were studied: branches per metre of stem, bark thickness, and height. If we were to carry this study out for real, we would certainly aim to use more replicates, and probably more variables as well. The results are:

	Branch density (m^{-1})	Bark thickness (cm)	Height (m)
GROUP 1			
Tree 1	2.6	1.8	12
Tree 2	3.6	2.5	13
Tree 3	4.8	5.1	15
Tree 4	4.1	4.2	14

	Branch density (m^{-1})	Bark thickness (cm)	Height (m)
GROUP 2			
Tree 1	5.9	2.6	13
Tree 2	3.0	1.7	11
Tree 3	5.5	5.1	14
Tree 4	4.0	2.0	12
GROUP 3			
Tree 1	4.0	6.4	13
Tree 2	2.5	2.7	11
Tree 3	1.9	5.4	12
Tree 4	3.5	7.2	14

Figure 10.1 shows *dotplots* for the three types of measurement. These are very similar to the frequency distributions shown in earlier chapters. They show how spread out the individual measurements in the samples are. If we were to compare either the branch densities *or* the bark thicknesses *or* the heights, using ordinary ANOVA we would find that we could not distinguish between the different populations. Looking at the dotplots, this is not surprising as there are large overlaps between the measurements from the three groups of trees. However, if we look at the same measurements on a 3D scatter graph (Figure 10.2), considering branch densities, bark thicknesses, and heights *together*, the points from group 1 do seem to be in a distinct cluster at the top left.

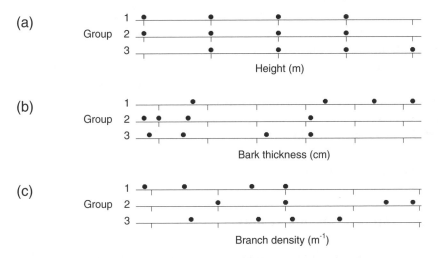

Figure 10.1   Dotplots for (a) height, (b) bark thickness and (c) branch density measurements. Samples of four individuals were taken from each of three populations (groups 1 to 3) and the height, bark thickness and branch density of each individual were measured. In a dotplot, each dot represents an individual measurement. Its position along the line gives the value measured: dots at the right-hand end of the line represent high values, dots at the left-hand end represent low values

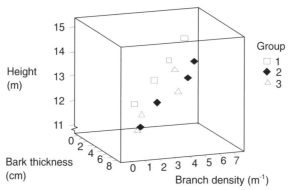

Figure 10.2   3D scatter graph of height, bark thickness and branch density measurements for samples of trees from three populations (groups 1 to 3)

Figure 10.3 shows a series of steps whereby the three types of measurement for each tree can be converted to a position along a line. If we now consider the individuals' positions along this line, the differences between the groups are fairly distinct (Figure 10.3(c)). There are several different mathematical approaches to MANOVA. One is effectively to fit a line like this and convert the measurements for each individual into a position along the line (equivalent to giving each individual a single value rather than three types of measurement). The populations are then compared using a procedure similar to ordinary ANOVA to compare their mean positions on the line. The line is fitted in such a way as to best discriminate between the different groups, so if a difference between the populations *can* be identified in this way, MANOVA will always find it. Other mathematical approaches are more difficult to represent diagrammatically but they produce similar results.

Because of the complexity of the calculations, MANOVA is almost exclusively carried out by computer, so it is necessary to ensure you have access to a statistical package which supports this. Data are usually arranged in a table where each row represents an individual member of a sample. One column is required for each factor; in this example there is only one factor, i.e. which *group* each tree comes from. One column is also required for each variable measured, i.e. one for branch density, one for bark thickness, and one for height.

10.5   Interpreting the results

A MANOVA test itself returns one or more *P*-values, equivalent to the lines in an ordinary ANOVA table (i.e. one *P*-value for each factor being investigated and one for each interaction). However, several different versions of the MANOVA test exist and some statistical packages give the *P*-values obtained by carrying out more than one of these. The Wilks, Lawley–Hotelling and Pillai tests are commonly used.

(a)

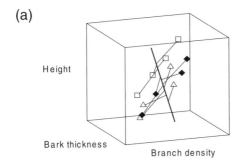

A line is fitted running across the three groups and each point is joined to this by a perpendicular line (in 3D)

(b)

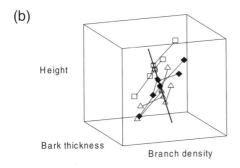

The position of each point along this new line is marked

(c)

Considering the positions of these points along the line, the different populations can then be compared

For example, we can see that trees from group 1 (white squares) produced points at one end of the line and trees from group 3 (white triangles) produced points at the other end

By converting the three types of measurement made on each tree to positions along a line like this, differences between groups are more clearly revealed

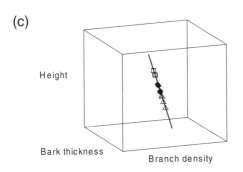

Figure 10.3   Schematic view of the stages involved in multivariate ANOVA

If the results of several versions of the test are given, they should all be similar, though they will not be identical because they use different mathematical approaches. The differences between them are quite technical but in normal work one can use any of them. Pillai's test is said to be least affected if the populations are not quite Normally distributed or have slightly unequal variances, so in the absence of any other reasons, this is a sensible choice. Here is an example of how the results might look from an experiment in which we have studied the effects of temperature, light and their interaction:

Effect of 'temperature'	
Wilks	$P = 0009$
Lawley–Hotelling	$P = 0.011$
Pillai	$P = 0.012$

Effect of 'light'	
Wilks	$P = 0.065$
Lawley–Hotelling	$P = 0.054$
Pillai	$P = 0.066$

Effect of 'temperature × light' interaction	
Wilks	$P = 0.041$
Lawley–Hotelling	$P = 0.044$
Pillai	$P = 0.043$

Note that ordinary univariate ANOVA would have given us a table with only three P-values, one for temperature, one for light, and one for the interaction between temperature and light. Here the computer has given three alternative P-values for *each* of these, depending on whose version of the test we want to use; other computer packages may give one, two or possibly more than three P-values for each effect, depending on which versions of the test they include.

We can use the results of any of these versions of the test and hopefully they will all show something similar. As with most other tests, by convention we take P-values of ≤ 0.05 to show a significant effect. From the above results, we could therefore conclude that there was an overall significant effect of temperature, no overall significant effect of light, but a significant interaction between temperature and light.

Unfortunately, but inevitably, some datasets give results that are just significant according to one version of the test and just not significant using another. It is legitimate to use whichever of the P-values you wish, as they are all bona fide types of MANOVA. However, to be totally objective about your results, it would be relevant to mention that other versions of the test would have led you to a different conclusion. This, of course, illustrates the drawback to having a convention about what is a significant difference. It leads us to try to classify results of statistical tests as black or white when really they are always shades of grey (Section 2.5).

A MANOVA test on the Scots pine data gives only one set of P-values as we are only studying one factor, which *group* the trees came from:

Wilks $P < 0.001$
Lawley–Hotelling $P < 0.001$
Pillai $P < 0.001$

We can conclude that there is a highly significant difference between at least two of the groups, whichever version of the test we choose. To determine which groups differ, we would need to compare the groups in pairs. This can also be done using MANOVA but leaving out the data for the third group in each case. Regardless of which version of the test we use, we find that for group 1 vs. group 2, $P = 0.002$; for group 2 vs. group 3, $P = 0.050$; and for group 1 vs. group 3, $P = 0.004$.

As we are making multiple comparisons – three in this case – we should adjust the P-value that we consider to show a significant effect using the Bonferroni method or an equivalent (Section 8.3). Hence only P-values of ≤ 0.017 (i.e. 0.05/3) should be considered significant. We can therefore conclude that group 1 was distinct from the other two but we have insufficient evidence to conclude that groups 2 and 3 were distinct from each other.

10.6   Further reading

If we are only comparing two groups, but comparing them in terms of several variables, there is such a thing as a multivariate t-test. This goes under various names, depending on whose version of it you use, one of which is *Hotelling's T^2 test*. Since MANOVA can also be used with only two groups and gives the same result, many statistical packages do not include a multivariate t-test as such. If a MANOVA test confirms for us that there are two or more distinct groups of individuals, *discriminant analysis* can be used to produce a decision rule to tell us for any further individuals we might measure, which group they are most likely to belong to.

11

Repeated Measures

11.1 Introduction

Studies of how some characteristic of a population changes with time will often involve measurements made on a number of occasions. However, we need to be clear about when an experiment or survey involves *repeated measures* in the statistical sense, because if it does we must take account of this in the statistical analysis. Repeated measures occur when our dataset involves measurements made on the *same* individuals or at the *same* points, on a number of occasions. Measurements made on the same population, but using a *different* sample of individuals or sampling points on each occasion, do not need to be treated as repeated measures.

This scenario involves repeated measures:

> Five soil cores are collected from a field and set up in a glasshouse so that water can be leached through them. Each week, for 6 weeks, 1 dm³ of water is leached through each core and the potassium concentration in the leachate is determined.

This scenario does not involve repeated measures:

> Thirty soil cores are collected from a field and set up in a glasshouse so that water can be leached through them. Each week, for 6 weeks, 1 dm³ of water is leached through each core. Each week the leachate is collected from five different cores and analysed for potassium concentration. The leachate samples are thus collected from *different* cores on each occasion.

You may find you have to study the above statements fairly carefully to see the difference. The distinction is, in fact, fairly subtle; very often we do need to think carefully to decide whether or not our experiment involves repeated measures.

Why is the distinction important? In the first scenario, consider what happens if we take a sample of five randomly located soil cores from the field and, by chance, they happen to be unusually high in potassium for that field. *Each week* when we collect the leachate, these five cores will give us readings that are high for that field.

In the second scenario we might also get a sample of five cores with unusually high potassium contents. However, these will give high readings for *one week only*. For the next week we will collect the leachate from five different cores. It is very unlikely that we will consistently get cores with potassium contents that are high for that field. Over the course of the experiment we would expect to get a fairly good representation of the field *on average* in the second scenario.

Since we can see that the likelihood of getting an unusual result is greater when we have used repeated measures, it makes sense that we need to make some allowance in the statistical analysis for these different approaches to collecting data.

There is nothing wrong with either of the two approaches above. There may be very good practical reasons for using one or the other. The important thing is to realize that the analysis of these two datasets differs. Suppose we want to compare potassium concentrations in leachate from two fields. Using the second scenario we have 30 *independent* measurements from each field, so these measurements would fulfil the requirement for ordinary ANOVA, that the data are independent measurements or observations (Section 8.2). Therefore we could compare the mean potassium concentrations in leachates collected at the different times and from different fields using ordinary two-way ANOVA.

Now consider what would happen if we analysed the data from the first scenario using ordinary two-way ANOVA. We could enter the numbers into a computer and carry out the analysis in the same way, but how would we tell the program that really we have only measured five cores, six times each, rather than 30 independent cores from each field, which is what the analysis is assuming? As discussed above, if we only measure five independent cores (six times each) our *overall* average result for that field will be *less reliable* than if we had used 30 independent cores. We therefore need to use an analysis which takes account of this. In this chapter we will look at some of the options for analysing repeated measures data.

Although repeated measures are usually datasets where something has been measured repeatedly in time, in the statistical sense we also include measurements that are *physically* linked together and therefore not independent. An example of this is sets of measurements made at different points along a river. Suppose we want to compare dissolved organic carbon (DOC) concentrations in rivers draining out of the west coast with those of rivers draining out of the east coast, and we also want to know how these vary with distance from the sea. Again let us consider two possible scenarios.

This scenario involves repeated measures:

In each of four rivers on each coast we make measurements of DOC at four distances from the sea (0, 1, 5 and 10 km). The concentrations are therefore measured at the four distances from the sea in the *same* eight rivers.

This scenario does not involve repeated measures:

> In four rivers on each coast we make measurements of DOC at a distance of 0 km from the sea. In different samples of four rivers on each coast we make measurements of DOC at a distance of 1 km from the sea. Similarly, we use different samples of four rivers on each coast to measure the concentrations at 5 and 10 km from the sea. The concentrations are therefore measured in *different* samples of rivers for each distance from the sea.

Like the example of the soil cores which might repeatedly give us high readings at successive times, if we happen to select a sample of four rivers with unusually high DOC concentrations at one distance from the sea, there is a good chance that they will have unusually high values at other distances too. The second scenario would give a more reliable overall average for each coast, but we would need to visit a total of 32 rivers rather than 8. Measuring the concentrations at different distances in the same 8 rivers might therefore be our only practical option. However, we need to recognize that if we do this then we are making repeated measurements and we will have to use methods such as those described in this chapter, rather than ordinary multiway ANOVA to analyse the data.

Experiments or surveys may be designed intentionally to involve repeated measures and in these cases we should try to make sure that a reasonable number of individuals are measured throughout. Problems usually only arise when people fail to recognize that their experimental design involves repeated measures until they come to analyse the data. Some other situations where repeated measures can occur are:

- Measurements taken at a series of depths down a soil profile, because these are physically linked together.

- Samples extracted at a series of time intervals from a set of flasks containing bacterial cultures.

In all of the above cases we *could* avoid making repeated measurements. In the case of the soil cores we could have used 30 cores instead of 5 and discarded the leachate from 25 each week. In the case of the rivers we could have measured at one distance from the sea in each of 32 rivers. You will see that there is a trade-off here. In exchange for collecting more *independent* measurements from our populations we have to do a lot more work, and in many cases this may be impractical.

In general, making repeated measurements on a small number of individuals will not reveal differences between populations as clearly as making the same number of measurements on a lot of independent individuals. However, if we are particularly interested in *changes over time* – more than the differences between the populations – then repeated measures turn out to be a very good way to carry out our study; see 'Comparisons between times' on page 167.

I will not introduce any new types of test in this chapter. All of the tests required have been described in Chapters 7 to 10. Instead I will look at ways of using them to analyse repeated measures data.

11.2   Methods for analysing repeated measures data

Consider the data shown in Figure 11.1. The experiment included two factors, carbon dioxide and watering treatment. Photosynthesis rates were measured on the same 12 leaves on each occasion. However, we are not interested in the individuals in the trial themselves; they are intended to represent how plants of this type in general would behave in these circumstances. If any of the leaves in our samples had unusually high (or low) photosynthesis rates on one occasion, the measurements from that leaf would probably be high (or low) on the other occasions as well, so we should allow for this when we try to draw conclusions about the populations they came from – we need to recognize that these are repeated measures data and analyse them accordingly. I am assuming here that the actual measurement process did not damage the leaf, otherwise succeeding measurements would not tell us anything useful, and I am also assuming that the different leaves were independent, i.e. attached to different plants.

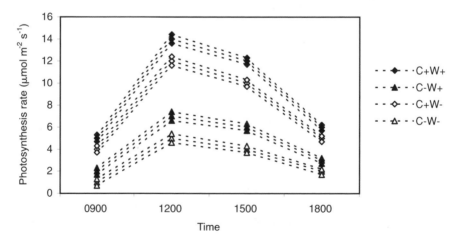

Figure 11.1   Measurements made on three replicate leaves in each of four CO_2 × water treatments. Treatments were combinations of high CO_2 (C+), or low CO_2 (C−) with high watering (W+) or low watering (W−). Each individual was measured on four occasions. The measurements at the different times are therefore 'linked' as indicated by the dotted lines

Comparisons between treatments at each time

Consider what happens if we take just the measurements made at 9:00am. We have 12 measurements made on *independent* leaves, so we can subject these values to ordinary two-way ANOVA (Chapter 8) to test for effects of CO_2, water and their interaction at that time. The fact that we later measured these leaves again does not affect the measurements we got at 9:00am. We can apply the same logic to the measurements made at 12:00noon. The fact that we had previously and subsequently measured these leaves does not affect the values at that time, so again we can use two-way ANOVA to test for treatment effects *at that time*. Similarly, we can use two-way ANOVA to test for treatment effects at 3:00pm and 6:00pm (Figure 11.2).

Comparisons between times

We can compare the mean photosynthesis rate at one time with the mean rate at any other time using a paired *t*-test (Chapter 7). This will tell us whether on average, considering all of the treatments together, photosynthesis rates differed between the two times (Figure 11.3).

We can also test whether the mean *change* between one time and another was different in the different treatments. First calculate the difference for *each* leaf, i.e.

(measurement at time 2) − (measurement at time 1)

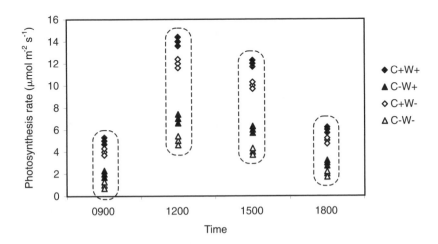

Figure 11.2   The measurements made at each time can be treated as four separate datasets and comparisons can be made between treatments at any particular time

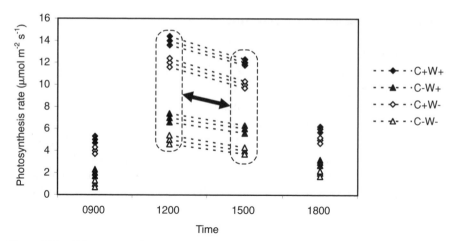

Figure 11.3 When comparing any two times with each other, each measurement at one time is paired with a measurement at the other time. The overall average photosynthesis rates at any two times can therefore be compared using a paired *t*-test

to get a set of 12 'changes in photosynthesis rate', and then carry out an ordinary two-way ANOVA for the effects of CO_2 and water on these values.

Analysis by summary

If we can attach a single number rather than a series of measurements to each individual in the experiment or survey, we can treat these values like ordinary rather than repeated measures data. For instance, in the example in Figure 11.1 we could calculate the overall mean for each leaf, i.e. the mean of the four measurements on each leaf. We would then have a set of 12 mean photosynthesis rates from 12 independent leaves. In effect, we are producing a set of figures that are *summaries* of the measurements for each leaf in the experiment. We could then use ordinary two-way ANOVA to compare these 'mean photosynthesis rates', in the different treatments. The downside to this is that we will only get 12 values from our 60 measurements to use in the analysis, so the comparison between treatments will not be very powerful.

We do not have to use the *mean* of the four measurements to summarize that leaf's behaviour. We could use another statistic such as the maximum value from each leaf, or the minimum value from each leaf. As long as we state clearly what we have done, and it is a logical measurement to take, the statistic we use to summarize each individual in the experiment is up to us. In experiments where there is a clear trend, such as a gradual increase in measurements over time, one option is to fit a regression line (Chapter 9) to the measurements for *each* individual. We could then take the *slope* of the line (Figure 9.5) to

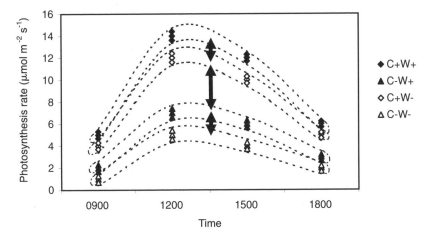

Figure 11.4   Treatments can be compared considering all of the measurement times together using MANOVA. This analysis takes into account that measurements in the dataset were made at different times and that the same individuals were measured on each occasion

summarize the measurements for that individual. This would be a measure of the 'average rate of change over the whole period' for each individual in the study. We could then compare these values in the different treatments using ordinary ANOVA.

Overall comparisons between treatments

We can use MANOVA (Chapter 10) to compare the treatments considering all of the measurement times together. An example of the use of MANOVA was given in Section 10.4, in which three variables were measured on each of a number of trees: branching density, bark thickness, and height. MANOVA was then used to compare the different groups on the basis of all these measurements considered at the same time. With repeated measures data, we can also consider the measurements made at the different times to be different variables.

On each individual we have measured its photosynthesis rate at 9:00am, its photosynthesis rate at 12:00noon, its photosynthesis rate at 3:00pm, and its photosynthesis rate at 6:00pm. Hence we can use MANOVA to test for overall differences between the populations, considering all of these measurements together (Figure 11.4). The test is simple to carry out and to interpret, assuming you have access to a statistical package with MANOVA.

The measurements from each time should be placed in a separate column so they will be treated as separate variables. Then, just as we can use ANOVA to compare between treatments at any *one* time (see above), we can use MANOVA to compare between treatments considering *all* the times combined. This is more likely to reveal differences between the treatments than just analysing the data from one time point, but often does not improve our chances as much as you might expect. After all, using the example of the photosynthesis measurements, we have still only observed 12 independent individuals in the experiment, however many times we make measurements on them.

11.3   Designing repeated measures experiments

Repeatedly measuring the same individuals is a good way of observing *changes* between times. When it comes to detecting differences *between treatments*, it is important to measure or observe as many independent individuals or independent sampling points as possible. Though we might set up an experiment with three replicates and measure each one every hour for a year, at the end we will still only have observed three individuals per treatment. We still cannot expect the analysis to show clear differences between the treatments on the basis of such a small sample.

A final consideration when designing an experiment with repeated measures is how far apart to make the measurements. If we have access to an automatic sampling system, it is often tempting to make measurements at very short intervals so that we do not miss anything. However, large datasets created in this way are often difficult to handle. There may even be too many data points for the computer package to handle. If we have to manipulate the data before analysing it, e.g. entering it into a spreadsheet in a suitable format, this might involve us in a lot of repetitive work. Also it is difficult to check that no errors have been introduced into the dataset.

Measuring individuals repeatedly is *most* useful for observing changes over time. It makes sense, therefore, to allow time intervals long enough for changes to take place, otherwise a lot of our analyses will show no significant differences between times. When using MANOVA to test for differences between *treatments*, there are also limitations with the degrees of freedom (Section 4.2). If you have too many time points in relation to the number of individuals measured at each time, the calculations required in this analysis cannot be done. It is sensible to check this using dummy data before collecting the real data (Section 4.2).

11.4   Further reading

Another common method for analysing repeated measures data, as an alternative to using MANOVA, is *analysis as a split-plot design*. This technique is described in many textbooks and, following their instructions, is reasonably straightforward. The downside to this method is that the assumptions it makes about the data are less likely to be met than those required by MANOVA, which is suitable for most datasets. You also need to know how to analyse a split-plot design on your particular statistical package. In its favour, analysis as a split-plot design is a more statistically powerful method, so it *might* reveal differences between groups even when the results from MANOVA are not significant. Another alternative approach to using MANOVA is *antedependence modelling* (Kenward 1987). This is also more powerful than MANOVA but is not widely available as a function on statistical packages.

12

Chi-square Tests

12.1 Introduction

The name 'chi-square' may also be written as χ^2, chi being a letter in the Greek alphabet. In either case it is pronounced 'kye-square'. Chi-square tests are used to analyse data that are *counts* of the numbers of individuals in different categories. Three of the main uses of chi-square tests are described in this chapter.

- *Goodness of fit tests*
  Used to test whether individuals are distributed among categories in accordance with some theory. For example, is the ratio of genotypes AA, AB, BB in a population 1:2:1?

- *Test for association between two factors*
  Used to test whether individuals that have one characteristic also tend to have another characteristic. For example, are people who are left-handed more likely to pass exams?

- *Comparison of proportions*
  Used to test whether the proportions of individuals with some characteristic are the same in different populations. For example, are the proportions of viable seeds the same in two seedlots?

Some computer packages that carry out chi-square tests leave you to do some of the calculations yourself, and may not include tests for goodness of fit at all. Fortunately, the calculations required for chi-square tests are straightforward to carry out using just a pocket calculator, so do not be put off using these tests just because you don't have the relevant software available.

12.2   Limitations and assumptions

The assumptions and requirements are the same for all of the chi-square tests described here; they are explained in Chapter 6:

- Random sampling

- Independent measurements or observations

In the case of chi-square tests, independent measurements or observations means that the individuals included in the counts must be acting independently. If we wanted to use a chi-square test to test whether sheep preferred particular areas for grazing, we might instinctively choose a time and count the number of sheep grazing on each area at that time. However, for our test to be meaningful, we would need to observe the preferences of a number of sheep acting independently. Since sheep do tend to follow each other around, this experimental design would probably not work. An alternative approach here would be to visit the area on a number of randomly selected days and count the number of *occasions* when there were sheep in each of the different areas. Assuming the sheep have time to move between the different areas in between visits, these observations would be independent of one another.

Chi-square tests make no assumptions about the distributions of values. Because of this they are among the least powerful tests. Consequently, it is necessary to use larger sample sizes than most other tests to have any chance of getting a significant result. However, if we only have to *count* individuals, rather than make measurements on them, it is often easy to include tens or even hundreds of individuals in the samples. The values used in the test must be actual numbers of individuals counted, not values converted to percentages. Suppose we had the following counts:

	First year	Second year
Males	20	35
Females	14	6

To present them as our results, we might choose to summarize the data as percentages:

	First year	Second year
Males	36%	64%
Females	70%	30%

However, we could not use the values 36, 64, 70 and 30 in the chi-square test.

The test takes account of the actual numbers observed (75 in total in this case). Using the percentages here would imply we had actually observed 200 individuals. If we only know the proportions and not the counts themselves, we cannot use a chi-square test. Of course, if you know the proportions and the total number of individuals counted, you can work out what the actual counts must have been and use those. If we are told that 300 individuals were observed and 20% were over 1 kg, we can calculate that 60 were over 1 kg and 240 were less than or equal to 1 kg, and use these values in a chi-square test.

The test involves calculating expected values (see below). None of these expected values should be less than 1, and no more than 20% of the expected values should be less than 5, or the test becomes inaccurate. If the expected values do not fit these criteria, you need either to collect more data, leave the categories with small expected values out of the analysis, or combine categories. For example, if we have categories for 10 different species of moth but we have counted only small numbers of some species, then we could try using genus or type of moth rather than species as the categories. These categories would include the counts of several species which might then all have large enough counts for us to use the test. Ensuring that none of the counts in our dataset is very small will help, but it is then necessary to recalculate and check again that the new expected values meet the requirements given above before using the test.

12.3   Goodness of fit test

Null hypothesis

The null hypothesis (H_0) and alternative hypothesis (H_1) for a goodness of fit test are as follows:

- H_0: the individuals are distributed in accordance with some specified proportions

- H_1: the individuals are not distributed in accordance with the specified proportions

Description of the test

An example of when we might use a goodness of fit test is:

It is suggested that dispersal of seeds from the edge of a plantation of trees follows an inverse square law (i.e. the number of seeds travelling x m is proportional to $1/x^2$). We want to test this theory using an isolated stand of trees on an area of moorland. Seeds will be collected in four 0.25 m² plots at each of five distances, on the downwind side of the plantation.

The total numbers of seeds counted at each distance were:

Distance (m)	5	10	15	25	50	
Seeds (observed value)	200	65	14	12	3	Total = 294

If seed dispersal follows the inverse square law, we would expect to find seed numbers at these distances in the following ratios:

Distance (m)	5	10	15	25	50	
Ratio of expected values						
As a fraction	1/25	1/100	1/225	1/625	1/2500	
As a decimal	0.04	0.01	0.0044	0.0016	0.0004	Total = 0.0564

We collected 294 seeds in total. If the inverse square law is true, we would expect 294 seeds to be distributed between the plots a follows:

At 5 m

$$\frac{0.04}{0.0564} \times 294 = 208.4$$

At 10 m

$$\frac{0.01}{0.0564} \times 294 = 52.1$$

At 15 m

$$\frac{0.0044}{0.0564} \times 294 = 23.1 \text{ etc.}$$

Distance (m)	5	10	15	25	50	
Expected number of seeds	208.4	52.1	23.1	8.3	2.1	Total = 294

These expected numbers of seeds are the 'expected values' referred to in Section 12.2. One of the values is less than 5 (i.e. 20% of the values are less than 5), which will not help the accuracy of the test (Section 12.2). It was stated in Section 12.2 that if more than 20% of expected values are less than 5, we cannot use the test, so according to that rule we are okay, just. Other texts give slightly different cut-off points. It is a question of accuracy not right or wrong. To be safer, and to illustrate the method, I will combine the last two categories so that we now have four categories, the last one being '25 and 50 m'. The observed count for '25 and 50 m' is 15, and of 294 seeds we would expect 10.4 to land at these distances.

We use the observed and expected numbers of seeds in each category to calculate a chi-square value as follows:

Distance (m)	5	10	15	25 and 50	
Observed values (O)	200	65	14	15	Total = 294
Expected values (E)	208.4	52.1	23.1	10.4	Total = 294
Difference ($O - E$)	−8.4	12.9	−9.1	4.6	
$(O - E)^2$	69.7	166.8	83.7	21.0	
$(O - E)^2/E$	0.334	3.202	3.616	2.016	Total = 9.168

The total of the values of $(O - E)^2/E$ is the chi-square value (i.e. $\chi^2 = 9.168$), which we can then look up in tables or using a computer function to get the P-value. To look up chi-square values, we also need to know the number of degrees of freedom (df). For a goodness of fit test,

$$\text{df} = \left(\begin{array}{c}\text{number of}\\\text{categories}\end{array}\right) - \left(\begin{array}{c}\text{number of pieces of information taken from the}\\\text{sample data to calculate the expected values}\end{array}\right)$$

In this case there are now four categories (the different distances, after combining 25 and 50 m). To work out the number of pieces of information taken from the data, imagine you were asked to calculate the expected values for the test. Starting with no data from the sample it can't be done. But if I gave you the *total* number of individuals (294), this would be all you would need to get the expected values in this case. We therefore need *one* piece of information from the sample data, so in this case df = 4 − 1 = 3. This is usually written as a subscript, i.e. χ^2_3. An example of when you might use more pieces of information from the sample data is if you wanted to test whether the data fitted a Normal distribution. In this case you would find that you needed to use the total number of individuals, the sample mean and the sample standard deviation (three pieces of information) to calculate the expected values.

Chi-square goodness of fit tests are often used to test the null hypothesis that something is distributed *equally* among several categories. In the above example we could, for instance, have tested the null hypothesis that the *same* number of seeds are dispersed in any direction from the stand trees, by using a set of 1 m^2 plots at, say, 10 m from the forest edge at the eight principal points of the compass. In this case the expected values for each plot would be

$$\frac{\text{total number of seeds collected}}{8}$$

Otherwise the test would be carried out in the same way.

Interpreting the results

Here are some examples of how to interpret the results of a chi-square goodness of fit test:

- If the result of a chi-square goodness of fit test is a *P*-value > 0.05, we could conclude:

  The results are consistent with the hypothesized distribution.

- If the result of a chi-square goodness of fit test is a *P*-value ≤0.05, we could conclude:

  We have significant evidence that the results are not consistent with the hypothesized distribution.

For the seed dispersal example, we got $\chi_3^2 = 9.168$, which gives us $P=0.027$. This is a significant result, so we can conclude that the dispersal of seeds was not consistent with the theory that seed dispersal follows an inverse square law. We should not be too blinded by this result. Although it suggests that seed dispersal did not follow an inverse square law as such, if you were to draw a graph of the data, you would see that it clearly followed something very close to an inverse square law.

12.4   Test for association between two factors

Null hypothesis

The null hypothesis (H_0) and alternative hypothesis (H_1) for a test for association between two factors are:

- H_0: there is no association between the two factors
- H_1: there is an association between the two factors

Description of the test

An example of when we might use a test for association is:

> Three species of moth are found on an island but their ecology is not well understood. We want to discover whether different species favour particular habitat types more than others. A number of traps are set up at randomly selected points in each of the different habitats and a record is made of the number of moths trapped of each species.
>
> We need to be able to assume that individual moths move around indepen-

dently and that the traps operate in such a way that trapping one moth does not affect our chances of trapping any other. This is more likely to be a problem in a survey of small mammals if we were using traps that could only hold one animal. In such a case we might be led to believe there were a lot more of one species present, when in reality they just found the traps quicker.

The results can be presented in a table as shown below, known as a *contingency table*. Contingency tables can have two or more rows and two or more columns. The two factors in this example are species (columns) and habitat (rows), and we are testing whether there is any evidence of an association between them. It does not matter which factor we put in columns and which in rows. The numbers of moths counted were as follows:

Observed counts (O)

	Species A	Species B	Species C	Row totals
Woodland	25	12	7	44
Field	6	12	2	20
Mountain	3	3	19	25
Column totals	34	27	28	89

Expected counts are obtained from

$$\frac{\text{row total} \times \text{column total}}{\text{grand total}}$$

Expected counts (E)

	Species A	Species B	Species C	Row totals
Woodland	$(44 \times 34)/89 = 16.81$	$(44 \times 27)/89 = 13.35$	$(44 \times 28)/89 = 13.84$	44
Field	$(20 \times 34)/89 = 7.64$	$(20 \times 27)/89 = 6.07$	$(20 \times 28)/89 = 6.29$	20
Mountain	$(25 \times 34)/89 = 9.55$	$(25 \times 27)/89 = 7.58$	$(25 \times 28)/89 = 7.87$	25
Column totals	34	27	28	89

None of the expected values is less than 5, so we can proceed with the test (Section 12.2). If we found that, say, some of the expected values for the field habitat had been too low, we could try combining the counts from the field and mountain categories so that we have two types of habitat instead of three (woodland and non-woodland), and then try calculating expected values for the new groups. Notice that the row and column totals are the same for the expected counts as for the observed counts. We can use this fact to check our calculations.

The formula for the expected values looks quite abstract but consider what it means. Overall the preference shown by the moths was 44 woodland, 20 field, and 25 mountain. If all species were attracted to the different habitats to the same extent (this is the null hypothesis, i.e. no association between species and preference for different habitats) we would expect to find species A moths distributed in this ratio too. That is, we would expect the 34 species A moths to be distributed in the ratio 44:20:25. Divide 34 in this ratio and you will get the expected values: 16.81:7.64:9.55, as above. The same logic applies to dividing up the counts for the other two species. The rest of the calculation is similar to the goodness of fit test. We calculate $(O-E)^2/E$ for each value in the table individually.

Differences $(O-E)$

	Species A	Species B	Species C
Woodland	8.19	−1.35	−6.84
Field	−1.64	5.93	−4.29
Mountain	−6.55	−4.58	11.13

$(O-E)^2$

	Species A	Species B	Species C
Woodland	67.09	1.82	46.82
Field	2.69	35.20	18.42
Mountain	42.91	21.02	123.98

$(O-E)^2/E$

	Species A	Species B	Species C
Woodland	3.99	0.14	3.38
Field	0.35	5.80	2.93
Mountain	4.49	2.77	15.76
Total			39.62

The total of the values of $(O-E)^2/E$ is the chi-square value (i.e. $\chi^2 = 39.62$), which we can then look up in tables or using a computer function to get the P-value. To look up chi-square values, we also need to know the number of degrees of freedom (df). In a test for association between two factors, we have

$$df = (\text{number of rows} - 1) \times (\text{number of columns} - 1)$$

In this case df $= (3 - 1) \times (3 - 1) = 4$; this is usually written as a subscript, i.e. χ^2_4.

Interpreting the results

Here are some examples of how to interpret the results of a chi-square test for association between two factors:

- If the result of a test for association between two factors is a P-value > 0.05, we could conclude:

  There is insufficient evidence to conclude that there is an association between the two factors.

- If the result of a test for association between two factors is a P-value ≤ 0.05, we could conclude:

  There is significant evidence of an association between the two factors.

Looking up $\chi_4^2 = 39.62$ gives us $P < 0.001$. This is a highly significant result, so we have strong evidence that the species of moth did differ in their preferences for different habitat types.

12.5 Comparing proportions

Several tests exist to compare proportions. An adaptation of the t-test is described in many textbooks. This is a more powerful approach than using a chi-square test but usually comes with the advice that the sample size should be fairly large and the proportions should not be too close to 0 or 1, or the test is not accurate. Thus it requires some judgement on the part of the user to decide whether the data meet these conditions.

Using a chi-square test to compare proportions, we have the same restrictions as other chi-square tests – none of the expected values should be less than 1, and no more than 20% of the expected values should be less than 5 – which at least is easy to check. The chi-square method also has the advantage that equality of several different proportions (rather than just two) can be tested at the same time, similar to the way we can compare several means at the same time using one-way ANOVA.

Null hypothesis

The null hypothesis (H_0) and alternative hypothesis (H_1) for comparing proportions are as follows:

- H_0: the proportion of individuals with a particular characteristic is the same in the different populations

- H_1: the proportion of individuals with a particular characteristic is not the same in the different populations

Description of the test

An example of when we might use a comparison of proportions is:

A farmer is concerned that a recent increase in the numbers of military aircraft carrying out training flights over his farm has affected the proportion of sheep having stillborn lambs in his flock of sheep. He has records from the previous year, so he is able to compare them with records from this year's lambing, which has just ended.

The counts of sheep having stillborn lambs were as follows.

Observed counts (O)

	Last year	This year	Row totals
Live	78	69	147
Stillborn	6	12	18
Column totals	84	81	165

We can think of the test as comparing the proportion of sheep having stillbirths last year (7.1%) with the proportion this year (14.8%); however, the test uses the actual numbers of animals recorded rather than the percentages.

In fact, the test boils down to a test for association between two factors, health of lambs and year. We are asking, Was a sheep more likely to have a stillborn lamb in one year than the other? The test is carried out in the same way as the test for association described above.

Expected counts are obtained from

$$\frac{\text{row total} \times \text{column total}}{\text{grand total}}$$

Expected counts (E)

	Last year	This year	Row totals
Live	$(147 \times 84)/165 = 74.84$	$(147 \times 81)/165 = 72.16$	147
Stillborn	$(18 \times 84)/165 = 9.16$	$(18 \times 81)/165 = 8.84$	18
Column totals	84	81	165

None of the expected values is less than 5, so we can proceed with the test (Section 12.2). If this assumption is not met, there are too few categories to try combining any, so we could not use a chi-square test. However, we might be able to use some other kind of test (Section 12.6). If the sample size is within our control, it is better to avoid this problem by using a large sample size. If it is possible to estimate the results you expect to get, try using these estimates to calculate expected values *before* collecting the data. If it looks as though any of these might be less than 5, try to use a larger sample size. The calculation is carried out as in the test for association.

Differences $(O - E)$

	Last year	This year
Live	3.16	−3.16
Stillborn	−3.16	3.16

$(O - E)^2$

	Last year	This year
Live	10.01	10.01
Stillborn	10.01	10.01

$(O - E)^2/E$

	Last year	This year	
Live	0.13	0.14	
Stillborn	1.09	1.13	
Total			2.50

The total of the values of $(O - E)^2/E$ is the chi-square value (i.e. $\chi^2 = 2.50$), which we can then look up in tables or using a computer function to get the *P*-value. As for a test for association between two factors, we have

$$\text{df} = (\text{number of rows} - 1) \times (\text{number of columns} - 1)$$

In this case df $= (2 - 1) \times (2 - 1) = 1$; this is usually written as a subscript, i.e. χ_1^2.

Interpreting the results

Here are some examples of how to interpret the results of a chi-square test for comparing proportions:

- If the result of a test comparing proportions is a P-value > 0.05, we could conclude:

   There is insufficient evidence to conclude that the proportion of individuals with the characteristic differed between any of the populations being compared.

- If the result of a test comparing proportions is a P-value ≤ 0.05, we could conclude:

   There is significant evidence that the proportion of individuals with the characteristic was not the same in all of the populations being compared.

Looking up $\chi_1^2 = 2.50$ gives us $P = 0.114$; this is a non-significant result. However, we must be careful about how we state our conclusion. In this case we know exactly how many sheep had stillbirths in that flock, so we know for certain that the proportion *in that flock* did differ between years. However, if we consider these sheep to be a representative sample of sheep in general in that area, we can make a conclusion about how stillbirth rates compared for sheep in general in that area in the two years. Our conclusion should then be that we have insufficient evidence here $(P = 0.114)$ to conclude that the difference between the years was due to anything other than natural variability. Note also that even if we had found a significant difference, this in itself would not have proven that it was due to the aircraft.

12.6   Further reading

Some authorities suggest applying *Yates' correction* whenever the chi-square value has only one degree of freedom, to improve the accuracy of the test. The correction involves subtracting 0.5 from the magnitude of each of the $(O - E)$ values before they are squared (i.e. -1.5 becomes -1.0, 1.5 becomes 1.0). These values would then be squared and used in the chi-square formula. This has the effect of reducing the calculated value of χ^2, hence slightly reducing the significance of the result. For the example of proportions of sheep having stillborn lambs, applying Yates' correction we would get $\chi_1^2 = 1.77$, $P=0.183$. Other authorities have suggested that this overcompensates and does not really improve the test (Sokal and Rohlf 1995).

   Fisher's exact test for 2×2 tables may be useful when some of the expected values are too small to use a chi-square test. G-tests are a direct alternative to chi-square tests for the types of examples given here, although chi-square tests are still more widely used. However, G-tests have the advantage that they can be extended to cover tests of association and interaction between three or more factors. For example, we can ask, In what ways are job type, sex, and type of transport used to go to work related?

13

Non-parametric Tests

13.1 Introduction

Chapter 7 described tests to compare two population means (paired and unpaired t-tests), Chapter 8 described tests to compare several population means (one-way ANOVA), and Chapter 9 described a test for association between two variables (correlation). These so-called *parametric* tests depend on the data coming from populations with Normal distributions, and Chapter 6 described ways of checking whether this assumption was met. If our data do not appear to come from populations with Normal distributions, often we can still use these tests by first transforming the data (Section 6.3). However, transforming the data has two possible disadvantages: (i) it may be more difficult to interpret exactly what the test is showing, (ii) it is not always possible to achieve a Normal distribution with a simple transformation.

If the data do not have a Normal distribution, in the cases of the tests listed above, an alternative is to use an equivalent *non-parametric* test. These are usually straightforward to interpret but less powerful, i.e. less likely to detect differences which are really there (Section 2.6). Transforming data and using non-parametric tests are equally valid options in many cases, but non-parametric tests may also be useful in situations where we cannot use one of the above tests because the data are really scores rather than values, (known as *ordinal data*) or because some values are off-scale (see below).

Non-parametric tests are also sometimes known as *distribution-free* tests. This term is somewhat misleading because many of them do have some assumptions about the distributions of values in the populations, though they are less stringent than for the tests described in Chapters 7 to 9; for example, a test may require that the shapes of the distributions we are comparing are the same as each other, though not necessarily Normal. The relevant assumptions are given along with each of the tests described below.

Note that although non-parametric tests do not assume the data come from a Normal distribution, they will still give correct results even if the data do come from a Normal distribution. These tests are therefore a safe option if you are unsure. Their only real disadvantage is that they are less powerful than *t*-tests, ANOVA and correlation.

Means and medians

Although I have described the tests as being alternatives to *t*-tests, ANOVA and correlation, strictly speaking, most non-parametric tests compare population *medians* rather than population means. Both the mean and median of a population can be thought of as describing a 'typical' member of the population, or at least *attempting* to describe a typical member of a population. We saw in Section 6.3 that for the following sample, the mean does not seem typical of the measurements at all:

30, 49, 74, 40, 63, 295, 60   mean = 87.3

The median is calculated by putting the measurements in rank order and taking the middle one, or the mean of the middle two if there is an even number of values. For the above example we would get

30, 40, 49, **60**, 63, 74, 295   median = 60

In this case the median seems more typical of the values. It is completely unaffected by how extreme the largest or smallest values are; consider what would happen if the largest value was 2950 instead of 295. In cases like this we therefore have two ways of comparing 'typical' members of populations: (i) transform the values so that the *mean* of the transformed values is fairly typical of the other transformed values, or (ii) use a non-parametric test to compare *median* values instead.

The median is also a useful way to define a 'typical' value in some cases where it is actually *impossible* to calculate the mean, for example when:

The dataset includes off-scale or approximate values

Suppose we have 1, 3, 5, 5, 8, 9, > 10, > 10, > 10, then the median is 8. The median has to be 8 because whatever the last three values actually are, we know that they are greater than the first six. Using a test that compares medians is the *only* possibility in this case. It is not possible to calculate the sample mean, so we could not use a *t*-test, ANOVA or correlation.

If you are carrying out a test on a computer, entering > 10 for the last three values will probably generate an error. However, provided you are using a test that uses the ranks rather than the actual values (including all of the tests described in this chapter) you can get round this by entering the last three values as 11; in fact, you could use any value greater than 10 here. Whatever the real values should be, they are then certain to end up with the highest ranks. Be sure to use the same value for all of the off-scale values. It might seem that once you have entered 11 for these values you could calculate the mean, so you could use a t-test. However, this would not give the correct result because we do not know what the values actually are. We are merely using this as a trick when entering the values into a computer to ensure that they are converted to the highest ranks by a non-parametric test.

The values are actually scores on some scoring system

An example is if we are interested in whether some plants come into leaf earlier than others. We might set up a visual scoring system and then give scores to a sample of plants each week (e.g. 1 = bud in overwintering state, 2 = bud swollen, 3 = bud partially open, 4 = bud almost fully open, 5 = leaves fully expanded). It will probably be necessary to enter the scores as numbers (1 to 5) into a computer, so it might seem that we could use a t-test to compare the means. However, it does not make sense to calculate the mean of these values because we would actually be calculating the mean of a set of descriptions.

When there is a clear order to the scores but the values themselves have no meaning, the data are called *ordinal data*. If our data are really ordinal data, using t-tests, ANOVA or correlation is not correct and a non-parametric test based on the ranks must be used. (Note that the 'scores' produced by PCA (Chapter 14) are not ordinal data and can be used in t-tests, ANOVA, and regression.)

Types of non-parametric test

Many non-parametric tests have been devised. Chi-square tests are also non-parametric tests but they have been given a separate chapter (Chapter 12) because they cover such a wide range of situations. This chapter looks at some of the other most useful non-parametric tests.

Alternatives to t-tests

- Mann–Whitney U-test
- Two-sample Kolmogorov–Smirnov test
- Two-sample sign test

Alternatives to ANOVA

- Kruskal–Wallis test

- Friedman's test

Alternative to correlation

- Spearman's rank correlation

Methods such as *t*-tests and ANOVA use some rather elegant mathematical theory and one can imagine that their development took a considerable amount of time. Most non-parametric tests, on the other hand, one can imagine being invented in a flash of inspiration while sitting in the bath. They all involve calculating a *test statistic* – similar to the *t*-value in a *t*-test or the *r*-value in a correlation test – which is then looked up in tables or on a computer to tell us how likely we would be to get this result if the null hypothesis of the test were true (i.e. the *P*-value). We can thank the inventors of the tests for going through the drudgery of actually working out these probabilities, but the basic ideas behind the tests themselves tend to be delightfully simple. A further attraction of non-parametric tests is that many of them have impressive-sounding names, as you can see.

13.2   Limitations and assumptions

The following assumptions and requirements apply to all of the tests described here; they are explained in Chapter 6:

- Random sampling

- Independent measurements or observations

For most of the tests, only very small sample sizes are required to be able to carry out the calculations. However, our aim is to draw conclusions about populations, based on samples taken from them, so the sample sizes we use should be sufficiently large to be representative. For this reason, sample sizes less than 6 are not recommended. The examples in this chapter use small sample sizes to demonstrate the methods. Remember, though, that these methods are less statistically powerful than tests such as *t*-tests and ANOVA. Therefore we should usually aim to use larger sample sizes to improve the likelihood of obtaining significant results if differences between the populations really do exist. Any additional assumptions are described below in the section for that test.

13.3   Mann–Whitney *U*-test

The Mann–Whitney *U*-test can be considered as an alternative to an unpaired *t*-test. It is equivalent to the Wilcoxon rank sum test (which is not the same as the Wilcoxon matched pairs signed ranks test that you may see described in some books). While Wilcoxon's version of the calculations is simpler to carry out and explain, and is often used, the name Mann–Whitney *U*-test has continued to be more commonly associated with the test – there's no justice!

Limitations and assumptions

The distributions of values in the two populations being compared should both be the same shape, but not necessarily Normal. The test could therefore be used to compare the medians of two populations similarly skewed to the right (Figure 6.3), but could not be used to compare medians of one population skewed to the left and one skewed to the right.

Null hypothesis

The null hypothesis (H_0) and alternative hypothesis (H_1) for a Mann–Whitney *U*-test are:

- H_0: there is no difference between the medians of the populations that the samples came from

- H_1: there is a difference between the medians of the populations that the samples came from

Description of the test

An example of when we might use a Mann–Whitney *U*-test is:

> To compare median soil penetration resistance in two fields by measuring at a number of randomly selected points in each field. Soil penetration resistance is the pressure required to push a metal cone through the soil. When measuring this, the force is usually applied by hand, so there is a limit to how high the measurements will go.

The soil penetration resistances in megapascals (MPa) from the two fields are:

Field 1	1.6, 1.6, 1.2, 2.4, 2.8, 1.4
Field 2	2.1, 2.9, >3.2, 2.4, 2.8, 2.0, 2.5

The value >3.2 was obtained because the operator was only able to push with a pressure up to 3.2 MPa. We cannot leave this out because that would artificially reduce the mean. We cannot therefore use a t-test to compare the fields.

To carry out the Mann–Whitney U-test, first place all the values in rank order but note which ones come from which field. Field 1 entries are shown in bold. The 3rd and 4th values are the same so both are ranked 3.5, etc.

Value	1.2	1.4	1.6	1.6	2.0	2.1	2.4	2.4	2.5	2.8	2.8	2.9	>3.2
Rank	1	2	3.5	3.5	5	6	7.5	7.5	9	10.5	10.5	12	13

Sum of ranks for field 1 = 28   Mean rank = 28/6 = 4.7
Sum of ranks for field 2 = 63   Mean rank = 63/7 = 9.0

The idea behind the test is that if both populations have the same distribution, we would expect the values from the two samples to be completely interspersed when we put them in rank order, hence we should get approximately equal mean ranks. The fact that we get a higher mean rank for field 2 is because the sample from field 2 contains more high values, suggesting that it might come from a population with generally higher values. To assign an accurate probability to this, we use the ranks calculated above to calculate the test statistic, U, and look this up in tables or carry out the test on a computer to get the appropriate P-value. The calculations for this example are given in Appendix A.

Interpreting the results

Here are some examples of how to interpret the results of a Mann–Whitney U-test:

- If the result of a Mann–Whitney U-test is a P-value > 0.05 and the sample medians are 1.9 MPa for field 1 and 1.1 MPa for field 2, we could conclude:

  There is no significant difference between the median soil penetration resistances of the two fields.

- If the result of a Mann–Whitney U-test is a P-value ≤ 0.05, and the sample medians are 1.9 MPa for field 1 and 1.1 MPa for field 2, we could conclude:

  Field 1 has a significantly higher median soil penetration resistance than field 2.

The result for the soil penetration resistance example given here is $P = 0.054$ (not significant), so we can say we have insufficient evidence to conclude that the fields had different median soil penetration resistances.

13.4   Two-sample Kolmogorov–Smirnov test

The two-sample Kolmogorov–Smirnov test can be considered as an alternative to an unpaired t-test but actually tests for *any* difference between the distributions that the two samples come from. A significant difference could mean that the population medians are different, the variances are different, or the shapes of the distributions are different. The test will not indicate which is the case.

Limitations and assumptions

There are *no* restrictions on the shapes of the distributions.

Null hypothesis

The null hypothesis (H_0) and alternative hypothesis (H_1) for a Kolmogorov–Smirnov test are:

- H_0: there is no difference between the distributions of the populations that the samples came from
- H_1: there is a difference between the distributions of the populations that the samples came from

Description of the test

An example of when we might use a Kolmogorov–Smirnov test is:

> To compare the distributions of time to germination for two batches of seeds. Samples of seeds from each batch are placed in suitable conditions for germination and the time in hours before each seed germinates is recorded.

The times to germination in hours are:

Batch 1	70, 75, 68, 85, 88, 70, 72
Batch 2	85, 44, 49, 64, 106, 90

Two seedlots may have the same *mean* time to germination but different *distributions*; for example, in one seedlot a few germinate very early and a few very late, in another the seeds all germinate at approximately the same time. These kinds of difference can be important because some of the seeds will be in the soil for longer than others before germinating, making them more prone to attack by fungi, insects or rodents. The Kolmogorov–Smirnov test can therefore alert us to any differences between the seedlots.

First place all the values in rank order. Values only appear once in this list, even if they appear more than once in the samples:

Values   44, 49, 64, 68, 70, 72, 75, 85, 88, 90, 106

Then, for each sample, calculate what proportion of measurements in that sample are less than or equal to each of the values. Let's go through that again slowly. In batch 1, none of the values are less than or equal to 44, none of the values are less than or equal to 49, none of the values are less than or equal to 64, $1/7$ ($= 0.14$) of the values are less than or equal to 68, $3/7$ ($= 0.43$) of the values are less than or equal to 70, etc. In batch 2, $1/6$ ($= 0.17$) of the values are less than or equal to 44, $2/6$ ($= 0.33$) of the values are less than or equal to 49, $3/6$ ($= 0.50$) of the values are less than or equal to 64, $3/6$ ($= 0.50$) of the values are less than or equal to 68, $3/6$ ($= 0.50$) of the values are less than or equal to 70, etc. We get the following table:

Value	44	49	64	68	70	72	75	85	88	90	106
Proportion of values in sample ≤ above value											
Batch 1	0.00	0.00	0.00	0.14	0.43	0.57	0.71	0.86	1.00	1.00	1.00
Batch 2	0.17	0.33	0.50	0.50	0.50	0.50	0.50	0.67	0.67	0.83	1.00

Next calculate the pairwise differences between these proportions, but ignore whether they are positive or negative, i.e. calculate $(0.00 - 0.17)$, $(0.00 - 0.33)$, $(0.00 - 0.50)$, $(0.14 - 0.50)$, etc. Once you have calculated all the differences, note the biggest difference, ignoring $+$ or $-$ signs.

Pairwise differences   0.17, 0.33, 0.50, 0.36, 0.07, 0.07, 0.21, 0.19, 0.33, 0.17, 0.00

The largest difference is 0.50. Let us think about what this actually means. 0% of the measurements in batch 1 are ≤ 64, whereas 50% of the measurements in batch 2 are ≤ 64, i.e. 50% more values in batch 2 were ≤ 64 – evidence that the samples do not come from populations with identical distributions. The largest difference, D, is the test statistic for this test. If D turns out to be only small, we will have little evidence to suggest that the distributions differ.

Since we can expect larger samples to yield a more reliable result, the sample size is also involved in calculating the *P*-value. To find the *P*-value from statistical tables, we need to multiply *D* by both of the sample sizes (in the above case $0.5 \times 7 \times 6 = 21$) and look up this value in a table specifically for the Kolmogorov–Smirnov test. Statistical packages will usually give you the *P*-value directly.

Interpreting the results

Here are some examples of how to interpret the results of a Kolmogorov–Smirnov test:

- If the result of a Kolmogorov–Smirnov test is a *P*-value > 0.05, we could conclude:

  There is no significant difference between the distributions of the populations from which the two samples came.

- If the result of a Kolmogorov–Smirnov test is a *P*-value ≤ 0.05, we could conclude:

  There is a significant difference between the distributions of the populations from which the two samples came.

The result for the example given here is $P = 0.4$ (not significant), so we have insufficient evidence to conclude that the distributions of times to germination differed in the two seedlots.

13.5   Two-sample sign test

This can be considered as an alternative to a paired *t*-test.

Limitations and assumptions

There are *no* restrictions on the shapes of distributions. Data can be actual measurements or counts, or scores on some scoring system (ordinal data). The sign test can be used in a wide range of situations but is not very powerful, so a large sample size is usually necessary. Sample sizes less than 6 will never produce a significant result.

Null hypothesis

The null hypothesis (H_0) and alternative hypothesis (H_1) for a two-sample sign test are:

- H_0: there is no difference between the medians of the populations that the samples came from

- H_1: there is a difference between the medians of the populations that the samples came from

Description of the test

An example of when we might use a two-sample sign test is:

> To compare median numbers of aphids per leaf on tomato plants in a greenhouse, before and after fumigation. A random sample of leaves is selected at the start of the trial (left attached to the plants), and counts of the numbers of aphids are carried out on each leaf before and after fumigation, i.e. each value in one sample is paired with a value in the other.

The sample values for numbers of aphids per leaf are:

Leaf number	1	2	3	4	5	6	7	8
Before fumigation	0	0	15	8	44	2	3	1
After fumigation	0	1	0	0	0	2	0	3

The usual choice for paired data is the paired t-test. However, if the pairwise differences do not appear to have a Normal distribution, we cannot use this without transforming the data. We might decide to use a two-sample sign test instead, either because we cannot achieve a Normal distribution by transforming the data, or because we would like a result that is easy to interpret.

For each pair, if the value in the second sample is higher then count one $+$; if the value in the second sample is lower then count one $-$; ignore ties. Finally, count the number of plus and minus signs.

Leaf number	1	2	3	4	5	6	7	8
Sign	(0)	+	−	−	−	(0)	−	+

Number of plus signs = 2
Number of minus signs = 4

If there is really no difference between before and after fumigation, we would expect to get approximately equal numbers of plus and minus signs, i.e. as many leaves gain as lose numbers of aphids.

To get an accurate P-value for the test, we need to use a set of tables for the *cumulative binomial distribution* and look up the P-value with

Number of successes $= 2$
Number of trials $= 6$
Success probability, $p = 0.5$

The number of successes is the number of plus signs *or* the number of minus signs, whichever is the smaller; the number of trials is the number of pluses *and* minuses, and the success probability is the probability of success in any trial – always use 0.5 for this test.

This will give us the result for a one-tailed test (Section 2.5). Unless you have good reason to use a one-tailed test, double the P-value to get the result for a two-tailed test, which is what we usually want. If you are doing the test on a computer, the test should give the correct P-value directly as part of the output.

Interpreting the results

Here are some examples of how to interpret the results of a sign test:

• If the result of a sign test is a P-value > 0.05 and the sample medians are 50 before and 2 after, we could conclude:

There is no significant difference between the medians of the populations from which the two samples came.

• If the result of a sign test is a P-value ≤ 0.05 and the sample medians are 50 before and 2 after, we could conclude:

The median value after is significantly lower than before.

The result for the aphid counts example is $P = 0.69$ (not significant), so we have insufficient evidence to conclude that fumigation had any effect on the median number of aphids per leaf. A small sample size was used here to illustrate the method. In practice we should use a much larger sample size for this test.

13.6 Kruskal–Wallis test

This can be considered as an alternative to one-way ANOVA.

Limitations and assumptions

The distributions of values in the populations being compared should all be the same shape, but not necessarily Normal. The test could therefore be used to compare the medians of several populations similarly skewed to the right (Figure 6.3), but it could not be used to compare medians of some populations skewed to the left and some skewed to the right.

Null hypothesis

The null hypothesis (H_0) and alternative hypothesis (H_1) for a Kruskal–Wallis test are:

- H_0: the populations all have the same median

- H_1: the populations do not all have the same median

Description of the test

An example of when we might use a Kruskal–Wallis test is:

> To compare median soil penetration resistance in three fields, by measuring at a number of randomly selected points in each field.

The results in MPa are:

Field 1	1.6, 1.6, 1.2, 2.4, 2.8, 1.4
Field 2	2.1, 2.9, >3.2, 2.4, 2.8, 2.0, 2.5
Field 3	1.1, 1.0, 0.9, 1.5, 1.2

This is essentially the same example as I used for the Mann–Whitney U-test, except that we are looking for any differences among several populations instead of a straight comparison between two. In fact, the Kruskal–Wallis test uses the same idea as the Mann–Whitney U-test. The values from all the samples are placed in rank order, noting which values come from which samples. The ranks related to each sample are then added up to give a set of *sums of ranks*. They are set out in the following table, where field 1 values are shown in bold, field 2 values in normal font, and field 3 values in italics:

Value	0.9	1.0	1.1	**1.2**	1.2	**1.4**	1.5	**1.6**	**1.6**
Rank	1	2	3	**4.5**	4.5	**6**	7	**8.5**	**8.5**
Value	2.0	2.1	**2.4**	2.4	2.5	**2.8**	2.8	2.9	>3.2
Rank	10	11	**12.5**	12.5	14	**15.5**	15.5	17	18

Sum of ranks for field 1 = 55.5   Mean rank = 55.5/6 = 9.2
Sum of ranks for field 2 = 98   Mean rank = 98/7 = 14.0
Sum of ranks for field 3 = 17.5   Mean rank = 17.5/5 = 3.5

If the samples all come from populations with the same distribution, we would expect the mean ranks for all the samples to be similar, as for the Mann–Whitney U-test. If any of the mean ranks are very different from one another, we have evidence that the population medians are not all the same. Assigning an accurate P-value to the test involves calculating a test statistic, H, comparable to U in the Mann–Whitney test, and looking this up in tables or using a computer. Special tables for the Kruskal–Wallis test exist for very small sample sizes, but for larger sample sizes the P-value can be obtained by looking up the value of H in chi-square tables. The calculations are given in Appendix A.

Interpreting the results

Here are some examples of how to interpret the results of a Kruskal–Wallis test:

- If the result of a Kruskal-Wallis test is a P-value > 0.05, we could conclude:

    There are no significant differences between the medians of any of the populations.

If the result of a Kruskal–Wallis test is a P-value ≤ 0.05, we could conclude:

    There is a significant difference between at least two of the population medians.

For the three-sample soil penetration resistance example $P = 0.004$, so we can conclude that at least one pair of population medians differed from each other.

As with a one-way ANOVA test, if we find that the medians are not all the same, we would often like to know which ones differ from each other significantly. One option to achieve this is to carry out a series of Mann–Whitney U-tests to compare each sample against each of the others. In this case we need to make a correction for multiple comparisons (Section 8.3). The Bonferroni method can be used here too. Suppose we are comparing three populations; first calculate the number of possible pairwise comparisons ($= \frac{1}{2} \times 3 \times (3 - 1) = 3$). Any pairwise comparison can therefore be considered significantly different if a Mann–Whitney U-test gives a P-value $\leq 0.05/3$, i.e. ≤ 0.017. This approach is okay for most purposes but it cannot be guaranteed to give results consistent with the Kruskal–Wallis test because the Kruskal–Wallis test uses information from all of the samples at the same time, whereas the Mann–Whitney test is only using information from two in any particular test. Other methods are described in some textbooks, equivalent to calculating a least significant difference, but these often require you to carry out some calculations by hand.

13.7   Friedman's test

Friedman's test can be considered as an alternative to one-way ANOVA but can be used to analyse randomized complete blocks designs (Section 4.4) where we have one factor of interest (e.g. fertilizer) and a blocking factor which we want to allow for but are not really interested in (e.g. fields). Hence the test will give us a result for whether there are any differences between fertilizer treatments, *allowing for* differences between fields.

Limitations and assumptions

The distributions of values in the populations being compared should all be the same shape, but not necessarily Normal. The data must be arranged in blocks (Section 4.4). For Friedman's test, each block must contain one, and not more than one, individual from each treatment group.

Null hypothesis

The null hypothesis (H_0) and alternative hypothesis (H_1) for Friedman's test are:

- H_0: there are no differences between the medians of any of the populations that the samples came from.

- H_1: there is a difference between the medians of at least two of the populations that the samples came from.

Description of the test

An example of when we might use Friedman's test is:

> To compare median soil matric suctions at different heights up a slope when the measurements have been made over a period of several days. Soil matric suction is

a measure of soil wetness but its values can vary over several orders of magn
and are unlikely to be Normally distributed. If it is not possible to make mea
ments in one day, we might carry out the sampling over several days, making
measurement at a different randomly selected point at each height on the slope on
each day. The effect of height on soil matric potential is the factor of interest here,
but we can use Friedman's test to allow for changes between days (blocks).

The results are:

	Soil matric suction (kPa) for the three slope positions			Rank		
	Bottom	Middle	Top		Bottom	Middle	Top
Day 1	6	5	8		2	1	3
Day 2	3	10	8		1	3	2
Day 3	4	15	26		1	2	3
Day 4	6	26	100		1	2	3
			Sum of ranks		5	8	11

The test works in a similar way to the Kruskal–Wallis test. Instead of ranking
all of the measurements together, we rank the measurements within each block.
In the table above, each day is a block and the measurements at the three
heights are ranked (1, 2, 3) within each block. If the matric suction is not really
affected by height, we would expect the rank orders we get within the blocks to
vary randomly from day to day, hence the overall sums of ranks should be
approximately the same.

Here there seems to be a trend; on most days the samples collected from the
higher positions on the slope have the highest matric suctions, hence higher
overall sums of ranks. This suggests that perhaps matric suction did depend on
position on the slope. To determine an accurate P-value, a test statistic, S, is
calculated from the sums of ranks and looked up in tables, or the test can be
carried out by computer. Special tables for Friedman's test exist for very small
sample sizes; for larger sample sizes the P-value can be obtained by looking up
the value of S in chi-square tables. The calculations are given in Appendix A.

Interpreting the results

Here are some examples of how to interpret the results of Friedman's test:

- If the result of a Friedman test is a P-value > 0.05, we could conclude:

  There is no significant difference between the medians of any of the populations.

- If the result of a Friedman test is a P-value ≤ 0.05, we could conclude:

There is a significant difference between at least two of the population medians (allowing for the blocking factor).

The result for the soil matric suctions example is $P = 0.105$ (not significant), so we have insufficient evidence from these samples to conclude that the median matric suction really differed between heights up the slope.

13.8   Spearman's rank correlation

This can be considered as an alternative to Pearson's product moment correlation, which is usually referred to simply as 'correlation'.

Limitations and assumptions

There are no assumptions about the shapes of distributions. The data can be measurements or counts, or ordinal data (Section 13.1).

Null hypothesis

The null hypothesis (H_0) and alternative hypothesis (H_1) for Spearman's rank correlation test are:

- H_0: there is no monotonic association between the two variables
- H_1: there is a monotonic association between the two variables

A monotonic association between X and Y means that increases in X are always associated with increases in Y, or increases in X are always associated with decreases in Y. Figure 13.1 shows six possible associations between X and Y: (a) to (d) show monotonic relationships, (e) and (f) show non-monotonic relationships. Unlike Pearson's correlation coefficient, a significant result may indicate a straight line association or a curved association, as long as it is monotonic.

Description of the test

An example of when we might use Spearman's rank correlation is:

To test for a relationship between atmospheric lead concentration in parts per

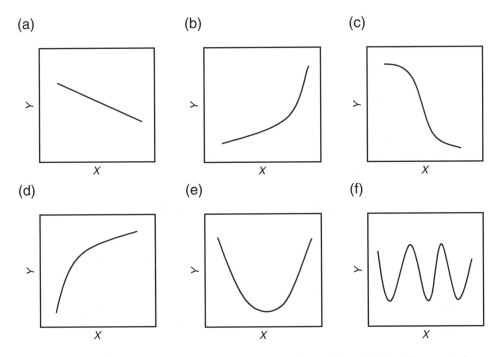

Figure 13.1 Different relationships between two variables, X and Y: (a–d) monotonic relationships, (e, f) non-monotonic relationships

billion (ppb) at the side of roads and size of road, where road size is recorded as a score (1 = minor, 2 = B class, 3 = A class, 4 = motorway).

The results are:

Road size score	1	3	2	3	1	4	4	2
Pb (ppb)	12	16	12	90	3	27	44	4

In this example one variable (lead concentration) is an actual measurement whereas the other is a score (ordinal data). Pearson's correlation requires *both* variables to be measurements or counts, so we cannot use it.

We start by ranking the measurements and scores *within* each variable, assigning 1 to the smallest, 2 to the next smallest, and so on. Any values that are tied are given the average of the ranks they occupy.

Road size score	1	3	2	3	1	4	4	2
Rank	1.5	5.5	3.5	5.5	1.5	7.5	7.5	3.5

Pb (ppb)	12	16	12	90	3	27	44	4
Rank	3.5	5	3.5	8	1	6	7	2

The calculations can be done in different ways, giving the same results. One way is to calculate the Pearson correlation coefficient (Section 9.3) but replacing each value in the dataset with its rank, as calculated above. Provided we can convert the data to ranks ourselves, Spearman's rank correlation coefficient, r_S, can therefore be calculated on computer packages which do not have a function specifically for it, simply by using the ordinary correlation function on the ranks.

The test is determining whether higher *ranks* of X are consistently associated with higher (or lower) *ranks* of Y. In the example we are testing whether higher lead concentrations are consistently associated with more (or less) major roads. Although the test statistic, r_S, is simply the r-value found by calculating Pearson's correlation coefficient using the ranks, its value must be looked up in special tables for Spearman's rank correlation coefficient (or using a computer version of Spearman's rank correlation test) to get the appropriate P-value. An alternative formula for calculating r_S is given in Appendix A.

Interpreting the results

Here are some examples of how to interpret the results of a Spearman's rank correlation test:

- If the result of a Spearman's rank correlation test is a P-value > 0.05 and the value of r_S is -0.78, we could conclude:

  There is no significant monotonic relationship between the X-values and the Y-values.

- If the result of a Spearman's rank correlation test is a P-value ≤ 0.05 and the value of r_S is -0.78, we could conclude:

  There is a significant negative relationship between the X-values and the Y-values.

The result for the lead concentrations example is $r_S = 0.82$ ($P = 0.025$), so we can conclude that there was a significant positive relationship between lead concentration and road size.

13.9   Further reading

If there are a lot of tied values in the Kruskal–Wallis test or Friedman's test, it is possible to apply a correction to improve the accuracy of the test. This has the effect of reducing the P-value but usually makes only a small difference. However, if a statistical package gives a P-value 'adjusted for ties', it makes sense to use it.

Many other non-parametric tests have been devised, though most rely on having access to a computer package that will perform the test, or a set of statistical tables specifically for that test. One test that might be useful but seems to have received little attention was proposed by Scheirer *et al.* (1976) as a non-parametric equivalent of two-way ANOVA. This allows both main effects and their interaction to be tested. The data are first converted to ranks and then ordinary two-way ANOVA is carried out on the ranks instead of the original values. A few simple calculations are then required to obtain ratios of some of the sums of squares given in the ANOVA table. The resulting ratios can be looked up in chi-square tables or using a computer to obtain the P-values. Although the function is not available in most computer packages, it should be fairly straightforward to carry out this analysis using the output from ordinary ANOVA. The method is described on page 446 of Sokal and Rohlf (1995), where it is described rather more accessibly than in the original paper.

14

Principal Component Analysis

14.1   Introduction

Principal component analysis (PCA) is one of a class of statistical techniques called *multivariate methods*. Multivariate methods are used to analyse datasets where several types of measurement or observation have been made on each individual in the sample. For instance, if we are studying gulls, we might measure the wingspan, wing width, overall length, beak length, foot length, head circumference, and mass of each individual. Note that the measurements can have different units (e.g. length in millimetres, mass in grams) but all of the different types of measurement are made on each individual bird.

PCA identifies which characteristics vary most between individuals. This might turn out to be one of the measurements we have taken (e.g. length or head circumference) but very often it turns out to be some characteristic derived from our measurements, such as the relationship between some of the length measurements and a bird's mass. For example, the analysis might suggest that the most variable characteristic is a score made up of

$$(0.3 \times mass) - (0.2 \times overall\ length) - (0.15 \times wingspan)$$

For any given mass, the greater the length measurements, the lower the bird's score will be on this scale and for given length measurements, the greater the bird's mass, the higher its score will be (scores on these systems can be positive or negative). This might be interpreted as a measure of the bird's 'density', a property that would be difficult to measure directly on live birds. The score is not a measurement of the bird's density per se but does give an indication of it. The most dense birds will have the highest scores, the least dense birds will have the lowest scores. Such scoring systems, which are part of the output of the analysis, are actually the *principal components* referred to in the name principal component analysis; see 'Loadings' on page 210.

Related to this, PCA can be used for data reduction. This means that instead of giving two measurements of length and one of mass for each bird, we might find that we could provide almost as much information about how birds differ by just giving each bird's 'density' score, leaving us with one set of values where previously we had three. With large datasets, it can be a considerable advantage if we reduce the number of values we have to handle. PCA will also tell us how much information we would be losing by reducing the dataset in this way; see 'Eigenvalues' on page 209.

Finding out which characteristics vary most between individuals might be an end in itself for some studies. However, PCA gives us several scoring systems (the different principal components) hence several scores for each individual, and these scores can be treated like new types of measurements to compare populations or test for correlations. For instance, we could test whether the birds' 'density' scores were correlated with how high up cliffs they nest, or we could carry out ANOVA on these scores to test whether different populations had different mean 'densities'; see 'Scores' on page 214.

We can also plot scatter graphs of the scores. This can sometimes reveal groupings of individuals that are not obvious in the original data; see 'Scatter graphs' on page 214.

Section 14.3 gives an example where PCA is used to identify groupings in a set of individuals where there is no prior information about different types. PCA is also often used when we already know something about groupings, e.g. when examples of a species of insect have been collected from four different countries. In this case the analysis might be used to find out whether insects from the different countries fall into distinct groups, which might lead us to propose that distinct subspecies exist.

PCA sometimes produces very interesting and exciting results. Groupings may emerge which one could not possibly have seen in the raw data. However, a big disadvantage is that often the results are not clear-cut. There may only be hints of groupings in the data, or the scoring systems may be difficult to interpret. There is no hypothesis test, no *P*-value, and no decision rule to tell you when you have a positive result. Some effort is therefore required on the part of the user to interpret the various scoring systems produced and any groupings that appear on the graphs. The analysis can also produce large tables of numbers that require some careful study, which may be off-putting. Hopefully though, after reading this chapter, you will have a good idea what you are looking for and not be too discouraged by the apparent volume of output.

PCA quite often produces somewhat disappointing or inconclusive results, so it is perhaps unwise to carry out an extensive study if this will be your only form of analysis at the end. However, the data will often also be suitable for cluster analysis (Chapter 15) and possibly also MANOVA (Chapter 10). It is common practice to carry out several types of analysis on multivariate data to fully understand how individuals vary, how they are grouped, and which types of measurements are closely related.

The dataset can also be considered as individual variables. In addition to carrying out PCA and other multivariate analyses, we might perhaps test for a correlation between overall length and wingspan ignoring the other variables, or analyse the beak lengths using one-way ANOVA to test whether birds from different sites differ in this respect.

14.2 Limitations and assumptions

Statistical *tests* often require certain assumptions to be met about the distributions of the data. This is because the probabilities (*P*-values) they calculate are based on these assumptions being met. In PCA we are not actually testing anything, we are just trying to identify sources of variability and groupings in the data. Therefore there are no limitations on the distributions of the data.

PCA uses correlations between variables (Chapter 9). However, it takes into account the correlations between several variables at the same time. As for ordinary correlation, the data should be measurements or counts of things. They should not be scores on an arbitrary scoring system (Section 13.1) and they cannot be coded versions of unordered categories (1 = square, 2 = circle, 3 = triangle, etc.).

As with all the methods in this book, if we are using measurements on a sample of individuals or individual points to infer something about a larger population, it is important that the sample is a fair reflection of the population. This is usually achieved by taking a random sample and ensuring that individuals within the sample are independent (Section 6.2).

If *any* of the measurements on an individual are missing, all of the other measurements for that individual must also be left out of the statistical analysis; in other words, that individual is excluded from the sample. Most statistical packages will do this automatically. However, it is worth bearing in mind that if we have 50 individuals in our sample and, say, the height measurements are missing for 40 of them, the analysis will be carried out on a sample size of 10. It might be better to leave out the height measurements completely and perform the analyses with the remaining variables for all 50 individuals.

14.3 Description of the method

There is no null hypothesis for basic PCA as described here.

An example of when we might use PCA is:

A consultant has been asked for advice about remediating the sites of 20 chemical factories abandoned during a war in a developing country. No records are available about what was manufactured at the different sites. He decides to start by carrying out chemical analyses on soil samples from each of the sites and measuring the concentration of 11 different substances (Subst1 to Subst11) in each:

- Subst1 to Subst3 are common inorganics

- Subst4 to Subst7 are common organics

- Subst8 to Subst10 are associated with the textile industry

- Subst11 is associated with the manufacture of explosives

At this stage he does not know how many different types of chemical factory there are, or which factories produced what.

The data for this example are given in Appendix B.

Calculating principal components without a computer is not a practical option, so it is important to make sure you have a statistical package that will do this. For most statistical packages the data should be arranged in columns with each variable, i.e. each *type* of measurement (length, height, etc.), in a separate column. Each row should then contain the measurements made on one individual.

There are a few decisions you need to make before carrying out the analysis, but unless you have good reason to do otherwise, you can follow the standard settings given below.

Should you use the covariance or correlation matrix?

Using the covariance matrix will apply more weighting to some variables than others, depending on the magnitude of the actual numbers (e.g. the diameter of a person's head in millimetres would be given more weighting than their height in metres, simply because the numbers are bigger). Using the correlation matrix applies equal weighting to all the different types of measurement in the analysis. If this means nothing to you, don't worry, just use the *correlation* matrix.

How many principal components should you calculate?

Principal components are the 'scoring systems' referred to above. They are produced in order of importance, so the first principal component is the scoring system that reflects the most variable characteristic of the sample. The second and subsequent principal components reflect gradually less and less variable characteristics. In practice you are unlikely to learn anything useful from more than the first three or four components. However, I would usually err on the side of caution and calculate around six components so that I don't miss anything interesting.

The maximum number of principal components you can calculate is equal to the number of variables – the number of types of measurement – in your dataset. So if you have only four variables you can only calculate up to four principal components. In effect, the program really always calculates all possible components and the decision you are making is just how many to display. Displaying more components will not affect any of the other components. The disadvantage of displaying a lot of components is only the amount of printout you will get. There can be *a lot*, and most of it will tell you nothing useful, so don't choose more than six components unless you have good reason.

What results should you store?

You are likely to want to store the *scores* for each of the principal components. You can use these scores to draw scatter graphs as described below.

You will need to specify which variables to include in your analysis, i.e. which columns of measurements. These variables will usually be *all* of those you have measured, but they do not have to be. Once you have specified which variables to include, and set up the other options as described, the results should be produced in a form similar to the example below (Section 14.4).

14.4   Interpreting the results

Eigenvalues

The different scoring systems produced are called *principal components* and each one has an eigenvalue. These are actually variances; we can think of them as saying, if we measure the population using this scoring system then there will be this much variation among the scores. Notice that they appear in order, with PC1 having the largest eigenvalue, PC2 the next largest, and so on.

	PC1	PC2	PC3	PC4	PC5	PC6
Eigenvalue	5.3603	2.8867	1.1247	0.6124	0.4396	0.2611
Proportion	0.487	0.262	0.102	0.056	0.040	0.024
Cumulative	0.487	0.750	0.852	0.908	0.948	0.971

	PC7	PC8	PC9	PC10	PC11
Eigenvalue	0.1792	0.0601	0.0461	0.0261	0.0037
Proportion	0.016	0.005	0.004	0.002	0.000
Cumulative	0.988	0.993	0.997	1.000	1.000

The lines to concentrate on here are not so much the eigenvalues themselves but the proportions and cumulative proportions of total variance they account for. We started out with 11 types of measurement for each factory (concentrations of Subst1 to Subst11). From these results, we can see that we could describe 48.7% (0.487) of the total variation between factories by giving just one score for each factory (its score on PC1), rather than all 11 types of measurement. We could describe 75.0% (0.750) of the total variation between factories by giving scores using both of the first two scoring systems (PC1 and PC2), and 85.2% (0.852) of the total variation using three scores (PC1, PC2 and PC3), i.e. three types of measurement rather than 11.

Although the technical meaning of eigenvalues might seem a bit obscure, a more down to earth interpretation is to say that, since 85% of the total variation between sites is concentrated in the first three principal components, we should be most interested in whatever types of measurement these three scoring systems are made up from.

In this example we have obtained a fortunate result where most of the variation is concentrated in the first few components. If the variation turns out to be spread fairly evenly over a lot of components, it can be difficult to interpret the results and it might be better to pursue some other type of analysis.

Scree plots

Scree plots are simply plots of the amount of variance (or eigenvalues) attributable to each of the different principal components, in order. The scree plot for the above results is shown in Figure 14.1.

Some people find it easier to decide how many components to focus on by using a scree plot. In this example there is little variance explained by the fourth and subsequent components, so I will concentrate on interpreting the first three. Unless there is a clear break in the slope of the curve, the choice of how many components to concentrate on is very subjective.

Loadings

In the terminology I have used up to now, the loadings are the scoring systems themselves, also known as *eigenvectors*. The loadings for the first six principal components are as follows:

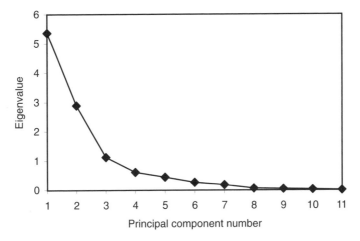

Figure 14.1   A scree plot for the principal components calculated for the chemical factories example

	PC1	PC2	PC3	PC4	PC5	PC6
Subst1	0.407	−0.033	−0.053	−0.066	0.036	0.512
Subst2	0.413	−0.038	−0.051	0.103	−0.274	0.088
Subst3	0.384	−0.106	−0.190	−0.322	−0.182	0.328
Subst4	−0.323	−0.003	−0.318	−0.004	−0.858	0.020
Subst5	−0.343	0.057	0.061	−0.680	0.037	−0.072
Subst6	−0.334	0.187	−0.004	0.588	0.038	0.409
Subst7	−0.375	0.119	−0.124	−0.236	0.200	0.623
Subst8	−0.093	−0.561	0.057	−0.033	0.052	0.153
Subst9	−0.130	−0.555	−0.003	0.066	0.024	−0.021
Subst10	−0.128	−0.557	−0.023	0.103	0.036	0.033
Subst11	0.014	−0.006	−0.913	0.046	0.327	−0.192

The loadings tell us how each scoring system is related to each of the original variables. Values close to zero indicate little relationship between scores on this system and the original variables. In this example we can see that scores on PC3 are closely related to Subst11 concentration. The loading is negative (−0.913) so we can say that factories with high scores on PC3 will have low concentrations of Subst11, and factories with low scores on PC3 (i.e. negative scores) will have high Subst11 concentrations.

Note that some programs produce tables of correlation coefficients rather than eigenvectors, but these may also be referred to as loadings. The interpretation of correlation coefficients is broadly similar, i.e. values in the table close to +1 or −1 indicate that a particular variable has a strong influence on the scores on a particular principal component. However, this type of 'loading' cannot be used directly to calculate the 'scores' as described below. You should therefore check which type of loadings your statistical package produces. We must now try to interpret the different principal components.

The first principal component (PC1)

Variables with loadings close to zero can be ignored. Unfortunately, it is rather subjective which ones are close enough to zero to ignore, but our aim here is to find a meaningful interpretation rather than a mathematical description of the data. This scoring system seems to be dominated by the first seven loadings and is approximately

$$0.41 \times [Subst1] + 0.41 \times [Subst2] + 0.38 \times [Subst3]$$
$$- 0.32 \times [Subst4] \ - 0.34 \times [Subst5] - 0.33 \times [Subst6] - 0.38 \times [Subst7]$$

where [Subst1] is the *standardized* concentration of Subst1, etc. (Box 14.1). The concentrations of Subst8 to Subst11 do not greatly affect the score on this component, so we can ignore them. PC1 therefore contrasts Subst1 to Subst3 (inorganics), which have positive coefficients in the scoring system, with Subst4 to Subst7 (organics), which have negative coefficients. The higher the concentration of inorganic substances found at the site, the higher the score; the higher the

---

Box 14.1   Standardizing measurements

Opting to use the correlation matrix in the analysis has the effect of making all the types of measurement equally important, regardless of the magnitude of the actual values. This is achieved by replacing each measurement in the analysis by a 'standardized' measurement:

$$\frac{standardized}{measurement} = \frac{measurement - mean \ of \ that \ type \ of \ measurement}{standard \ deviation \ of \ that \ type \ of \ measurement}$$

The different types of measurement are more correctly called variables. In the factories example, Subst1 concentration is one variable, Subst2 concentration is another variable, and so on. For Subst1 concentration, we have a mean of 4.48 and a standard deviation of 1.018 (Appendix B). The Subst1 concentration for factory 1 is 3.3, so

$$standardized \ Subst1 \ concentration \ for \ factory \ 1 = \frac{3.3 - 4.48}{1.018} = - 1.16$$

A standardized value can be calculated for each of the measurements in the original dataset in a similar way.

   After standardizing, each variable will have a standard deviation of 1 and a mean of 0. This procedure for standardizing variables is not unique to PCA. It is used in many other statistical methods also.

concentration of organic substances, the lower the score. Hence PC1 can be considered to be a measure of how inorganic or organic are the residues at the site.

The second principal component (PC2)

Ignoring variables with loadings close to zero, this scoring system is approximately

$$- 0.56 \times [\text{Subst8}] - 0.55 \times [\text{Subst9}] - 0.56 \times [\text{Subst10}]$$

The concentrations of Subst1 to Subst7 and the concentration of Subst11 do not greatly affect the score on this component. Subst8, Subst9 and Subst10 were the chemicals associated with textile manufacture. Hence PC2 might be considered to be a measure of how likely factories are to have been associated with the textile industry. Since the loadings all happen to be negative values, the lower the score, the more likely a factory is to have been involved with the textile industry.

The third principal component (PC3)

Ignoring variables with loadings close to zero, this scoring system is approximately

$$- 0.32 \times [\text{Subst4}] - 0.91 \times [\text{Subst11}]$$

The concentrations of Subst1 to Subst3 and the concentrations of Subst5 to Subst10 do not greatly affect the score on this component. Without more information about the chemistry involved, it is not obvious what the combination of Subst4 and Subst11 signifies, but this scoring system is dominated by the concentration of Subst11. Hence PC3 can be considered to be a measure of how likely a factory is to have been involved in arms manufacture. Since the loading for Subst11 is negative, a low score indicates higher likelihood of being involved with arms manufacture.

One might attempt to interpret the other principal components in a similar way but in general this becomes progressively more difficult as we get to higher-numbered components. Also, as we are interested in differentiating between types of factory, we should concentrate our efforts on the components that include most of the variance.

One other common scenario is to obtain a component where all the loadings are approximately the same, either all positive or all negative. If we had obtained such a component in this example, we could interpret it as a general index of pollution due to chemical manufacture. Increases in any of the substances would have had a similar effect on this score.

Scores

Box 14.2 explains how the scores are derived from the loadings. The scores are therefore 'measurements' of how each factory rates on the different scoring systems given by the principal components:

Factory	PC1	PC2	PC3	PC4	PC5	PC6
1	–1.922	2.062	0.566	0.594	–0.864	–0.285
2	2.620	1.230	0.592	0.433	–1.003	–0.340
3	–2.917	1.225	–1.507	–0.516	0.093	0.231
4	2.165	–2.030	0.793	0.957	0.074	0.484
5	1.794	–2.002	–1.228	–0.186	0.301	–0.150
6	1.386	0.771	–1.371	–0.754	0.146	0.302
7	–1.196	2.785	1.112	0.917	1.212	0.456
8	–2.020	–0.771	–1.302	0.600	0.321	–0.412
9	2.661	0.763	–1.434	0.199	–0.085	–0.378
10	–2.310	–0.268	0.485	–1.212	–0.914	–0.747
11	1.982	0.720	–1.294	0.310	0.461	0.245
12	–1.405	0.032	1.087	0.965	0.319	–0.595
13	–2.358	–1.476	0.765	0.181	–0.138	–0.315
14	1.815	–2.220	0.531	0.709	–0.898	0.141
15	–2.184	–0.855	–1.437	0.579	0.071	–0.275
16	–1.653	1.363	1.005	–0.342	0.861	0.092
17	2.970	1.807	0.930	–1.739	0.316	–0.530
18	2.621	–2.762	1.161	–0.602	1.005	–0.102
19	1.713	1.895	0.157	0.001	–0.988	1.111
20	–3.760	–2.270	0.388	–1.094	–0.290	1.068

We can use these to gain information about particular factories. We can see that factories 2, 4, 9, 17 and 18 have particularly high scores on PC1, i.e. high concentrations of inorganic chemicals compared to organics. Factories 7 and 18 have the highest and lowest scores respectively for likelihood of being involved with the textile industry (PC2), and so on. Most statistical packages will save the scores for you when you carry out the analysis. These scores are equivalent to measurements made on the individuals. They can be analysed like other types of measurement using tests such as t-tests, ANOVA or regression.

Scatter graphs

Plotting scatter graphs of the scores given by different principal components

Box 14.2   Calculating scores on principal components

This explanation assumes you have used the correlation matrix in the analysis, as most people do. If you have used the covariance matrix, the procedure for calculating the scores is the same except that you use the actual measurements instead of the standardized measurements (see below). The procedure described also assumes that the loadings given by your statistical package are eigenvectors and not correlation coefficients.

Each factory has a score on each principal component. These can be calculated from the loadings and the standardized substance concentrations for that factory (Box 14.1). For factory 1, the score on PC1 is calculated as follows:

$0.407 \times$ standardized Subst1 concentration for factory 1	$0.407 \times (-1.16) = -0.472$
$0.413 \times$ standardized Subst2 concentration for factory 1	$0.413 \times (-0.47) = -0.196$
$0.384 \times$ standardized Subst3 concentration for factory 1	$0.384 \times (-1.16) = -0.444$
$-0.323 \times$ standardized Subst4 concentration for factory 1	$-0.323 \times \quad 1.12 = -0.360$
$\vdots$	$\vdots$
$0.014 \times$ standardized Subst11 concentration for factory 1	$0.014 \times (-0.76) = -0.010$
	Total $= -1.922$

So the score for factory 1 on PC1 is -1.922. Scores for other factories, and on other principal components, can be obtained in the same way. In practice most statistical programs will calculate the scores for you.

may help to reveal groupings of individuals that are not obvious from the original data. To understand this, let us first consider plotting some measurements not involving PCA. Suppose a potato grower buys a load of seed potatoes from a new supplier one year. At the end of the year there seems to be a lot more variation in his crop than usual and he wonders if the seed potatoes he bought were actually not all of the same quality. He takes a sample of 40 plants at random and measures three variables on each: the dry mass of the above ground part, the dry mass of potatoes produced, and the starch concentration in the potatoes.

He starts by looking at frequency distributions of the above ground mass, potato mass, and starch concentration (Figure 14.2). There are no clear groupings here. Next he tries plotting a scatter graph of above ground mass vs. potato mass (Figure 14.3). Again, there are no obvious groups here. Finally he tries starch concentration vs. potato mass (Figure 14.4). This time we can see there are apparently two different groups of plants. This seems to suggest that there were actually two different lots of seed potatoes mixed together. He might question his supplier about this.

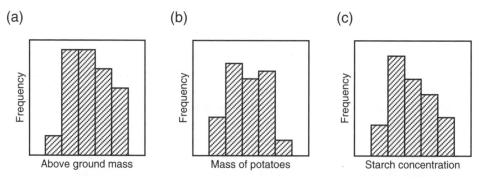

(a)   (b)   (c)

Figure 14.2   Frequency distributions for the three variables measured on each potato plant:
(a) above ground mass, (b) mass of potatoes, (c) starch concentration in the potatoes

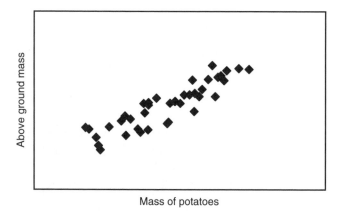

Figure 14.3   Scatter graph of above ground mass vs. mass of potatoes produced by individual plants

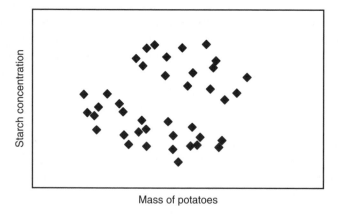

Figure 14.4   Scatter graph of starch concentration in the potatoes vs. mass of potatoes produced by individual plants

Of course, if there really are no distinct groupings, whichever variables we plot against each other will not show this kind of pattern. Yet why was it that plotting above ground mass vs. potato mass did not reveal these groupings but plotting starch concentration vs. potato mass did? This was largely because above ground mass and potato mass were highly correlated, i.e. the plot in Figure 14.3 is almost a straight line. On the other hand, when measurements are fairly uncorrelated, as potato mass and starch concentration were in this example, any groupings that exist are more likely to stand out.

We could draw scatter graphs for different pairs of substance concentration measurements for the factories example in a similar way, but with 11 types of measurement there would be 55 graphs to draw. Also we can only make use of two types of measurement at a time. However, an alternative is to use our new measurements here – the scores on each of the principal components – and plot scatter graphs of them.

These have the advantage that each type of score is making use of information about a number of substance concentrations rather than just one. Also, because of the way they are derived, the different types of score have the useful property that they are all uncorrelated with each other; just what we are looking for to reveal groupings in the data. Because most of the variation between factories is expressed in their scores on PC1, PC2 and PC3, we will concentrate on these three principal components (Figure 14.5).

Various groupings of factories emerge. In this case the computer program has labelled the points with the number of each factory. Otherwise we could identify which factory was which by reading the scores off the graph and checking back on the table of scores. Factories at the bottom of Figure 14.5(a) are most likely to be related to the textile industry. How far to the right they appear indicates the amount of inorganic substances relative to the amount of organic substances found.

Figure 14.5(b) suggests there are four groups of factories. Particularly concerning are those at the bottom with highly negative scores on PC3. This indicates a possible involvement with arms manufacture. These can be divided into those where relatively large amounts of inorganic chemicals were found (5, 6, 9 and 11) and those where relatively large amounts of organic chemicals were found (3, 8 and 15), which might tell us something about the types of weapons. We should note, however, that factory 5 was also among those quite likely to be involved with the textile trade, so we need to consider whether Subst11 has other uses too.

This is a fictitious example, so we cannot analyse the chemistry in too much detail. However, we can see that in a situation like this, PCA could be a useful tool to identify groups of individuals in a complex dataset, which would help us to plan a more detailed sampling strategy for further studies.

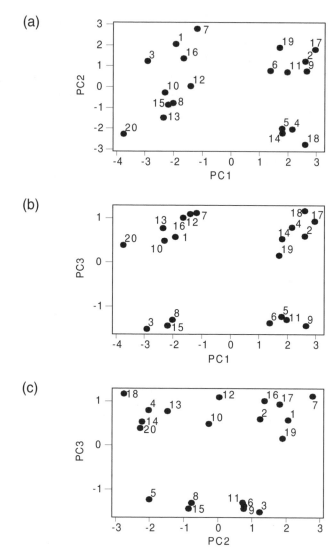

Figure 14.5 Scatter graphs of scores on principal components 1, 2 and 3. The numbers next to individual points are the numbers assigned to different factories in the original dataset

14.5   Further reading

In the example above we identified several scoring systems, such as PC1, which was interpreted as the degree of organic/inorganic pollution at the site. However, our main interest was in using these scoring systems to study differences between factories, and groupings of factories. If your *main* aim is to look for underlying factors (similar to the organic/inorganic scores in the example), that are not directly observable in your dataset, the closely related technique of *factor analysis* may be more suitable. For example, if we have a set of measurements for how mice behave in different situations (speed of running away, mass of cheese stolen, etc.) and we suspect there may actually be some underlying but unmeasurable factor causing the differences between individuals (such as courage or greed), we could use factor analysis to try to identify this.

15

Cluster Analysis

15.1 Introduction

Cluster analysis is one of the group of statistical techniques known as multivariate methods. These are so called because they work on datasets in which we have more than one type of measurement, or variable, made on each individual. Hence multivariate methods are used to reveal differences, similarities or relationships between individuals, considering *several* different types of measurement at the same time (e.g. height, weight, speed and age). Datasets used for MANOVA (Chapter 10) and PCA (Chapter 14) are usually also suitable for analysing by cluster analysis.

Suppose we have collected soil samples from five sites and they have the following characteristics:

Site	pH	Dry bulk density (g cm^{-3})	Sand content (%)	Clay content (%)
1	5.3	1.2	14	22
2	5.6	1.1	18	6
3	5.4	1.6	12	18
4	4.4	1.5	26	40
5	4.6	1.1	25	9

We might ask the question, Which sites are most similar to each other? If we consider only pH, sites 1, 2 and 3 seem to form one group, while sites 4 and 5 form another. If we consider only bulk density, sites 1, 2 and 5 seem to form one group, while sites 3 and 4 form another. If we want to decide which sites were most similar overall then it becomes more difficult. We can use cluster analysis to decide which of the sites are most similar to each other considering all of the properties at once.

The most commonly used type of cluster analysis starts by considering every individual to be a group. Then it decides which two individuals are most similar and groups them together. Then it decides which individuals or groups are next most similar and these are grouped. Groups are thus built up from subgroups, and so on. This is called *agglomerative hierarchical clustering*.

Cluster analysis can be used in two ways:

- *Clustering observations*: this is deciding which *members of the sample* are most similar to each other.

- *Clustering variables*: this is deciding which of the different *types of measurement* we have made are most similar, i.e. which variables are the most highly correlated.

In either case the result is usually presented as a *dendrogram*, which gives a clear indication of which individuals or variables are most similar, and which form groups distinct from other individuals or variables.

Cluster analysis does not involve testing any hypothesis, hence no P-value is produced from which we can make a clear decision. Interpretation of the results therefore requires some thought on the part of the user. The calculations behind the analysis are complex and will almost always be carried out by computer, so although the output might give us a clear picture of which individuals are most similar, what it really means to be 'similar' in this respect is often not so clear. Most of the time a general understanding is all that is required. However, there are different ways of measuring similarity and our choice of measure can influence the groupings we find. Understanding why can take some considerable thought (see the sections on linkage method and distance measure in Section 15.3).

15.2   Limitations and assumptions

The assumptions quoted for statistical tests are required mainly because we are aiming to give a probability, a P-value, of something happening by chance if these assumptions are true. Since cluster analysis does not include any hypothesis testing, the assumptions are rather more relaxed. There are no restrictions on the shapes of distributions of values for basic cluster analysis as described here.

If we want to use the results of a cluster analysis to infer something about a larger population, it is important that the sample is representative of the population as a whole. This is usually achieved by taking a random sample (Section 6.3). However, there may also be occasions when we want to use cluster analysis and we have access to a whole population, in which case the issue of sampling does not arise. An example of this would be a study to look for similarities in the combinations of chemicals discharged from paper mills in the UK.

No missing values are allowed in the dataset (Section 3.4). If there are missing values, you can either delete all the measurements for the individuals concerned (i.e. one or more whole rows of data) or delete the variables concerned (i.e. one or more whole columns of data) to remove them.

The analysis involves calculating 'distances' between individuals in the dataset (see below). Data that are measurements or counts are therefore suitable; for example, it makes sense to say that two individuals with heights 1 cm and 3 cm are *as alike* as two individuals with heights 7 cm and 9 cm. Data that are scores on an arbitrary scoring system (Section 13.1) or coded versions of unordered categories (1 = square, 2 = circle, 3 = triangle, etc.) are not suitable for ordinary cluster analysis as described in this chapter. However, some computer packages do include versions of cluster analysis suitable for these types of data.

15.3   Clustering observations

For ordinary cluster analysis as described here, there is no null hypothesis or alternative hypothesis. We are not actually testing any statement; we are simply looking for patterns in the data.

Description of the method

To illustrate the similarities and differences between the results of PCA and clustering observations analysis, I have used the dataset for the factories example in Section 14.3. Cluster analysis is an alternative way to look for any natural groupings within the list of 20 factories. The data are given in Appendix B.

There are a few decisions you need to make before carrying out the analysis:

Linkage method

The analysis starts by deciding which two individuals are most similar. These are then joined together. The joining of individuals or clusters is called a *fusion*. The next fusion may be joining together two other individuals or joining another individual to this cluster, or later in the analysis, joining together two clusters. Each fusion is the joining together of the next most similar pair of individuals or clusters. The question arises, How do you decide which is the most similar cluster to an individual or another cluster. Figure 15.1 shows three common methods.

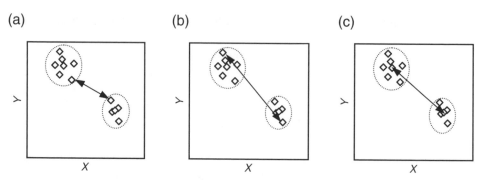

Figure 15.1 Different linkage methods for defining the distance between two clusters: (a) single linkage or nearest neighbour, (b) complete linkage or furthest neighbour, (c) centroid linkage. The coordinate of the centroid is given by the mean of the X-coordinates and the mean of the Y-coordinates of the points in the cluster. When calculating the distances between clusters, the linkage method defines *which* distances are measured. The distance itself can then be measured in several different ways (Figure 15.2)

With single linkage, only the distances between the closest pairs of individuals are considered. With complete linkage, only the distances between the individuals furthest apart are considered. With centroid linkage, the distances between the centroids of the clusters are considered. Computer packages may offer other types of linkage. In general, there is no right or wrong method and the usual advice is to try several in the hope that one or more give clear and meaningful results.

Distance measure

The linkage method tells us *which* distances are going to be measured and compared to find the next most similar individuals or clusters, but we can also *measure* these distances in different ways. Figure 15.2 shows two ways of defining the 'distance' between two points when only two variables have been measured. This can be extended to three-dimensional space, which we could represent on a 3D graph, with a little effort. In fact, though it becomes difficult to visualize, mathematically these methods can be extended to any number of dimensions. Therefore we can calculate the distance between two points when any number of variables have been measured for each point. Other types of distance measure may be offered by computer packages, and like linkage type, in general it is reasonable to try any of them.

Standardizing the variables

Suppose the number of petals on the flowers of a plant is always 4, 5 or 6, while plant heights vary from 10 cm to 2000 cm; number of petals will have little effect

(a)   (b)

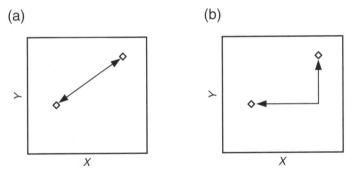

Figure 15.2 Different measures of distance between two points: (a) Euclidean distance, (b) Manhattan or city block distance

on the distance between individuals, which will be largely just a reflection of their heights. Usually we want all of the variables measured to have equal importance when deciding which individuals are most similar, so we need to *standardize* the variables (Box 14.1). This is usually available as an option on computer packages. If it is not, check the manual to see if this is done by default.

The analysis itself will almost invariably be carried out by computer. For most computer packages the data need to be arranged with each type of measurement (i.e. each variable) in a separate column, and each individual in the sample on a separate line. The program then goes through the data, successively joining the most similar individuals or clusters until all individuals have been joined into one large cluster. The results are usually displayed graphically.

Interpreting the results

Cluster analysis of the factories data produced the dendrogram in Figure 15.3. Other choices of linkage type and distance measure would have produced slightly different groupings.

In this case the axis has been labelled 'similarity', with the lowest values (least similarity) at the top of the axis. Without changing the diagram, the axis could also be (and sometimes is) labelled 'distance', with the largest distances at the top of the axis.

From Figure 15.3 we might discern that factories 8 and 15 are particularly similar (90% similar), while factory 20 had only 46% similarity to any of the other clusters. Perhaps more usefully, the analysis suggests there are four distinct clusters that are no more than 33% similar to each other: (i) 4, 5, 14, and 18; (ii) 2, 6, 9, 11, 17 and 19; (iii) 1, 7 and 16; (iv) 3, 8, 10, 12, 13, 15 and 20. We might therefore choose to investigate the factories further, considering them to be of four different types.

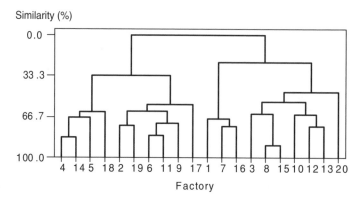

Figure 15.3   Dendrogram produced by cluster analysis of the factories data using complete linkage and Euclidean distance, with the variables standardized

Notice that groupings suggested by cluster analysis correspond approximately with the groupings produced by PCA in Figure 14.5(a). Compared with this figure, it is interesting to note that the cluster analysis has placed factories 1, 7 and 16 into a distinct group, whereas factory 3 is in a different group. Differences like this arise because cluster analysis is using *all* of the variables at once, while Figure 14.5(a) is making use of the information in principal components 1 and 2 only. These largely ignore the concentration of Subst11, which is given by principal component 3.

When we consider the scores on component 3 (Figure 14.5(b)) it can be seen that factory 3 is much more similar to factories 8 and 15, than it is to factories 1, 7 and 16. We do not have to include all the variables we have measured in the cluster analysis. We could simply leave some out and run the analysis again, in which case other groupings may appear. The analysis itself can make no allowance for whether some of the variables have more practical importance than others.

Cluster analysis is very straightforward to use and produces results that broadly agree with PCA. In most cases it is probably easier to observe groupings in cluster analysis than PCA. However, as in the example used here, it may be easier to understand what the groupings represent if they have been produced by PCA. Using a combination of cluster analysis and PCA is therefore often a good approach.

15.4   Clustering variables

As with clustering observations, there is no null hypothesis or alternative hypothesis for clustering variables. We are simply looking for patterns in the data.

Description of the method

Consider the scenario for the factories example in Section 14.3. The concentrations of 11 specified substances were measured at each factory because these were known to be associated with certain types of chemical manufacture. To illustrate the similarities and differences between the results of PCA and clustering variables analysis, I have used the same dataset again. We would therefore expect this analysis simply to show that the substances are grouped according to what type of chemical process they were used for. Not very useful here but it might be a useful type of analysis if, for example, we had detected 11 unidentified but commonly occurring substances in soil samples, and we wanted to know whether some combinations of these were more common than others.

Clustering variables is essentially about grouping together variables that are highly correlated. One can get some idea about which variables are similar simply by calculating the correlation coefficients, r, between each variable and each of the others (Section 9.3). These can conveniently be presented in a matrix, called a *correlation matrix*. In fact, clustering variables can then be carried out fairly easily by systematically picking out pairings of variables which are the most highly correlated, then the next most highly correlated, and so on. However, computer packages can conveniently do this for us and present the results in a dendrogram similar to that obtained in clustering observations.

Interpreting the results

Figure 15.4 shows the dendrogram obtained from a cluster analysis of the variables in the factories example from Section 14.3.

The figure shows four distinct clusters: Subst1 to Subst3, Subst4 to Subst7,

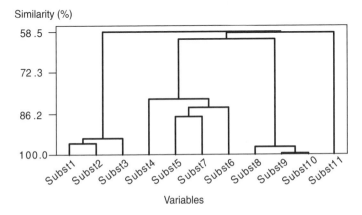

Figure 15.4   Dendrogram produced by cluster analysis of variables for the factories data using single linkage

Subst8 to Subst 10, and Subst11. In the example these correspond to inorganics, organics, substances from the textile industry, and a substance used in explosives manufacture. Since Subst9 and Subst10 have very high similarity – they are very highly correlated – we might conclude that there is no real need to measure both. If one is high then the other will also be high, and vice versa.

In the example used here, we might have expected the variables to be clustered in this way. In real-life situations, clustering variables can be a useful way to discover which traits or groups of traits tend to vary together in a population.

15.5   Further reading

Other distance measures and linkage methods exist apart from those described here, and they might be useful in some situations. There are several other forms of clustering, including *non-hierarchical* methods in which we attempt to divide the individuals into a specified number of groups, and *divisive* methods which start by considering all the individuals to be in one cluster and then look for the best way to split the cluster, and so on. Divisive methods work in the opposite way to the agglomerative methods in this chapter, which start by considering the individuals to be separate groups and gradually join them into larger clusters. Methods also exist to cluster individuals on the basis of qualitative data (colour of eyes, colour of hair, etc.) by determining how many characteristics they have in common.

Various methods are available to help decide how many clusters there really are, but these methods can be tedious to carry out if they are not available as functions on a computer. Also they only provide the best solution according to some mathematical formula. Studying the dendrogram and using a little scientific insight will probably give a much better interpretation of the way individuals or variables are really grouped.

APPENDICES

Appendix A
Calculations for Statistical Tests

s_1 = standard deviation of sample 1 (Section 2.1 and Box 2.1)

s_1^2 = variance of sample 1 (Section 2.1)

n_1 = sample size for sample 1 (Section 4.2)

$\bar{x}_1$ = mean of sample 1 (Section 2.1)

$|x|$ = the absolute value of x (e.g. $|3.2| = 3.2$, $|-1.4| = 1.4$)

df = degrees of freedom (Section 4.2)

Σx_1 = all of the values in sample 1 are added together

Σx_1^2 = each value in sample 1 is squared then these squares are added together

$(\Sigma x_1)^2$ = all of the values in sample 1 are added together and then the total of these is squared

Σxy = each pair of x and y values are multiplied together then these values are added together

$(\Sigma xy)^2$ = each pair of x and y values are multiplied together then these values are added together, and then the total of these is squared

As most people now obtain P-values directly from a statistical computing package, I have not included statistical tables in this book. They are, however, widely available in the back of other statistics textbooks and books of statistical tables (e.g. Neave 1978). Unfortunately, they vary somewhat in how they are laid out. I have therefore included small extracts along with the tests below so that you can check that you are using the tables correctly. Do not be put off by very minor differences in the values, as different sets of tables may have been produced in slightly different ways.

In general you need to check whether the test statistic you have calculated (e.g. a t-value or an F-value) is greater than or smaller than the tabulated values and from this you derive the P-value. This may tell you that P lies somewhere between two values or that P is outside some range in one direction or the other. Usually there is no real need to know P exactly. Except in the case of the Mann-Whitney U-test, the larger the value of the test statistic, the smaller the value of P, and the more significant the result.

Summary statistics (page 11)

Sample	$8, 12, 10, 7, 7, 11, 8 \quad n = 7 \quad \Sigma x = 63 \quad (\Sigma x)^2 = 3969 \quad \Sigma x^2 = 591$

Mean:
$$\bar{x} = \frac{\Sigma x}{n} = \frac{63}{7} = 9.0$$

Variance:
$$s^2 = \frac{\Sigma x^2 - \frac{(\Sigma x)^2}{n}}{n-1} = \frac{591 - \frac{3969}{7}}{7-1} = 4.0$$

Standard deviation: $s = \sqrt{s^2} = \sqrt{4.0} = 2.0$

Standard error: $\text{s.e.} = s/\sqrt{n} = 2.0/\sqrt{7} = 0.76$

95% CI:
$$95\% \text{ CI} = \bar{x} \pm t_{n-1,0.05} s/\sqrt{n} = 9.0 \pm 2.447 \times 2.0/\sqrt{7}$$
$$= 7.15 \text{ to } 10.85$$

where $t_{n-1,0.05}$ is a value obtained from t-tables with $(n-1)$ degrees of freedom and $P = 0.05$.

Unpaired t-test (page 95)

Sample 1	100	107	115	96	102	110	109	103	99	106	$\bar{x}_1 = 104.7$	$s_1^2 = 33.344$
Sample 2	95	90	102	100	98	90	100	95	88	105	$\bar{x}_2 = 96.3$	$s_2^2 = 32.233$

1. Calculate a pooled estimate of variance

$$s_p^2 = \frac{(n_1 - 1)s_1^2 + (n_2 - 1)s_2^2}{n_1 + n_2 - 2} = \frac{(9 \times 33.344) + (9 \times 32.233)}{10 + 10 - 2} = 32.789$$

2. Calculate the t-value with $\text{df} = (n_1 + n_2 - 2)$

$$t_{n_1 + n_2 - 2} = \frac{|\bar{x}_1 - \bar{x}_2|}{\sqrt{s_p^2\left(\dfrac{1}{n_1} + \dfrac{1}{n_2}\right)}} = \frac{|104.7 - 96.3|}{\sqrt{32.789\left(\dfrac{1}{10} + \dfrac{1}{10}\right)}} = 3.280$$

3. From t-tables

for $P = 0.05$, $t_{18} = 2.10$
for $P = 0.01$, $t_{18} = 2.88$
for $P = 0.001$, $t_{18} = 3.92$

Therefore, in this case P lies between 0.01 and 0.001, a significant result. Using a computer we find that $P = 0.004$.

Paired t-test (page 97)

Sample 1	400	20	24	95	228	116	65	112	35	45	81	197	$\bar{x}_1 = 118.2$
Sample 2	345	8	29	81	204	140	36	75	47	5	65	187	$\bar{x}_2 = 101.8$

1. Calculate pairwise differences (x_d).

   55, 12, -5, 14, 24, -24, 29, 37, -12, 40, 16, 10

2. Calculate mean and variance of pairwise differences in the same way that you would calculate the mean and variance of a sample.

   $\bar{x}_d = 16.333$ $s_d^2 = 513.70$

3. Calculate the t-value with df $= n_d - 1$, where n_d is the number of pairwise differences.

$$t_{n_d - 1} = \frac{|\bar{x}_d|}{\sqrt{\dfrac{s_d^2}{n_d}}} = \frac{|16.333|}{\sqrt{\dfrac{513.70}{12}}} = 2.496$$

4. From t tables

for $P = 0.05$, $t_{11} = 2.20$
for $P = 0.01$, $t_{11} = 3.11$
for $P = 0.001$, $t_{11} = 4.44$

Therefore, in this case P lies between 0.05 and 0.01, a significant result. Using a computer we find that $P = 0.030$.

One-sample t-test (page 99)

Sample	3.01	2.98	2.97	3.00	3.03	3.10	2.94	2.91	2.99	2.99
$n = 10$		$\bar{x} = 2.99$		$s^2 = 0.002\,618$			$\mu_0 = 3$ (hypothesized mean)			

1. Calculate the t-value with df $= (n - 1)$.

$$t_{n-1} = \frac{|\bar{x} - \mu_0|}{\sqrt{\dfrac{s^2}{n}}} = \frac{|2.99 - 3|}{\sqrt{\dfrac{0.002618}{10}}} = 0.494$$

2. From t-tables

$$\begin{aligned}
&\text{for } P = 0.05, && t_9 = 2.26 \\
&\text{for } P = 0.01, && t_9 = 3.25 \\
&\text{for } P = 0.001, && t_9 = 4.78
\end{aligned}$$

Therefore, in this case P is greater than 0.05, a non-significant result. Using a computer we find that $P = 0.63$.

Least significant difference (page 101)

From the example used for the unpaired t-test, above $s_p^2 = 32.789$, $n_1 = 10$, $n_2 = 10$.

$$\text{LSD} = \sqrt{s_p^2\left(\frac{1}{n_1} + \frac{1}{n_2}\right)} \times t_{n_1+n_2-2,0.05} = \sqrt{32.789\left(\frac{1}{10} + \frac{1}{10}\right)} \times 2.101 = 5.38$$

In the above example the difference between the sample means was $104.7 - 96.3 = 8.4$. This is greater than the LSD of 5.38, hence the test produced a significant result.

F-test (page 103)

Sample 1	100	107	115	96	102	110	109	103	99	106	$\bar{x}_1 = 104.7$ $s_1^2 = 33.344$
Sample 2	95	90	102	100	98	90	100	95	88	105	$\bar{x}_2 = 96.3$ $s_2^2 = 32.233$

1. Calculate F with $df = (n_1 - 1)$ and $df = (n_2 - 1)$.

$$F_{n_1-1,n_2-1} = \frac{s_1^2}{s_2^2} = \frac{33.344}{32.233} = 1.034$$

Always divide the larger variance by the smaller; if s_2^2 had been larger we would have calculated $F_{n_2-1,n_1-1} = s_2^2/s_1^2$.

2. From F-tables

for $P = 0.025$, $F_{9,9} = 4.03$
for $P = 0.005$, $F_{9,9} = 6.54$

These are one-tailed probabilities, so we need to multiply them by 2 (Section 7.4), i.e. we can consider these to be the F-values for $P = 0.05$ and $P = 0.01$. Therefore, in this case P is greater than 0.05, a non-significant result. Using a computer we find that $P = 0.96$.

Note that $F_{2,6}$ is not the same as $F_{6,2}$, and similarly for other values, so if n_1 and n_2 are not equal, it is important to get them the correct way round when looking up F-values in tables. So that you can check your use of the tables, for a one-tailed P-value of 0.025, $F_{2,6} = 7.26$, $F_{6,2} = 39.3$.

One-way ANOVA (page 111)

Sample 1	1.14, 1.22, 0.98	$n_1 = 3$	$(\Sigma x_1)^2 = 11.156$
Sample 2	1.01, 1,16, 1.25	$n_2 = 3$	$(\Sigma x_2)^2 = 11.696$
Sample 3	0.95, 0.99, 1.15	$n_3 = 3$	$(\Sigma x_3)^2 = 9.548$
Sample 4	1.51, 1.37, 1.30	$n_4 = 3$	$(\Sigma x_4)^2 = 17.472$
Sample 5	1.19, 1.39, 1.24	$n_5 = 3$	$(\Sigma x_5)^2 = 14.592$
Totals	$\Sigma x_T^2 = 21.614$	$n_T = 15$	$(\Sigma x_T)^2 = 318.622$

$(\Sigma x_T)^2$ is calculated by adding *all* 15 sample values together then squaring the total

Σx_T^2 is calculated by squaring each of the 15 values then adding these squares together

1. Calculate sums of squares.
   Total sum of squares

$$SS_T = \Sigma x_T^2 - \frac{(\Sigma x_T)^2}{n_T} = 21.614 - \frac{318.622}{15} = 0.3730$$

Between groups sum of squares

$$SS_G = \frac{(\Sigma x_1)^2}{n_1} + \frac{(\Sigma x_2)^2}{n_2} + \frac{(\Sigma x_3)^2}{n_3} + \frac{(\Sigma x_4)^2}{n_4} + \frac{(\Sigma x_5)^2}{n_5} - \frac{(\Sigma x_T)^2}{n_T}$$

$$= \frac{11.156}{3} + \frac{11.696}{3} + \frac{9.548}{3} + \frac{17.472}{3} + \frac{14.592}{3} - \frac{318.622}{15} = 0.2468$$

Within groups (or residual) sum of squares

$$SS_R = SS_T - SS_G = 0.3730 - 0.2468 = 0.1262$$

2. Section 8.3 explains how to set out an ANOVA table of these results. The calculation of degrees of freedom, mean squares, and the F-value are given in the notes below the table. We get $F_{4,10} = 4.889$. The F-value takes its degrees of freedom from the df for 'between groups' and the df for 'within groups', in that order.

3. From F-tables

for $P = 0.05$, $F_{4,10} = 3.48$
for $P = 0.01$, $F_{4,10} = 5.99$
for $P = 0.001$, $F_{4,10} = 11.28$

Although ANOVA is a two-sided test – it tests for any differences between means, not just differences in a particular direction – there is no need to multiply the P-values obtained from F-tables by 2 when using ANOVA. Therefore, in this case P lies between 0.05 and 0.01, a significant result. Using a computer we find that $P = 0.019$.

Two-way ANOVA (page 119)

The example involves two factors, island (factor A) and beach type (factor B) Factor A has four levels: Tinkewinke (level *a*), Larlar (level *b*), Dipsae (level *c*), and Sun Island (level *d*). Factor B has two levels: sandy (level 1) and shingly (level 2). It does not matter which we call factor A and which factor B.

Note that $(\Sigma x_T)^2$ is calculated by adding *all* 24 sample values together then squaring the total. Σx_T^2 is calculated by squaring each of the 24 values then adding these squares together.

	Factor A				
Factor B	Level a	Level b	Level c	Level d	Row totals
Level 1	1.5, 1.7, 1.0	3.6, 3.9, 3.1	2.6, 3.2, 2.1	2.0, 1.5, 1.9	
	$n_{a1} = 3$	$n_{b1} = 3$	$n_{c1} = 3$	$n_{d1} = 3$	$n_1 = 12$
	$(\Sigma x_{a1})^2 = 17.64$	$(\Sigma x_{b1})^2 = 112.36$	$(\Sigma x_{c1})^2 = 62.41$	$(\Sigma x_{d1})^2 = 29.16$	$(\Sigma x_1)^2 = 789.6$
Level 2	1.9, 1.2, 1.5	2.1, 2.5, 2.2	1.9, 2.3, 2.5	2.6, 2.1, 2.0	
	$n_{a2} = 3$	$n_{b2} = 3$	$n_{c2} = 3$	$n_{d2} = 3$	$n_2 = 12$
	$(\Sigma x_{a2})^2 = 21.16$	$(\Sigma x_{b2})^2 = 46.24$	$(\Sigma x_{c2})^2 = 44.89$	$(\Sigma x_{d2})^2 = 44.89$	$(\Sigma x_2)^2 = 615.0$
Column	$n_a = 6$	$n_b = 6$	$n_c = 6$	$n_d = 6$	$n_T = 24$
totals	$(\Sigma x_a)^2 = 77.44$	$(\Sigma x_b)^2 = 302.76$	$(\Sigma x_c)^2 = 213.16$	$(\Sigma x_d)^2 = 146.41$	$(\Sigma x_T)^2 = 2798.4$
					$\Sigma x_T^2 = 128.3$

1. Calculate sums of squares.

Total sum of squares

$$SS_T = \Sigma x_T^2 - \frac{(\Sigma x_T)^2}{n_T} = 128.3 - \frac{2798.4}{24} = 11.710$$

Factor A sum of squares

$$SS_A = \frac{(\Sigma x_a)^2}{n_a} + \frac{(\Sigma x_b)^2}{n_b} + \frac{(\Sigma x_c)^2}{n_c} + \frac{(\Sigma x_d)^2}{n_d} - \frac{(\Sigma x_T)^2}{n_T}$$

$$= \frac{77.44}{6} + \frac{302.76}{6} + \frac{213.16}{6} + \frac{146.41}{6} - \frac{2798.4}{24} = 6.695$$

Factor B sum of squares

$$SS_B = \frac{(\Sigma x_1)^2}{n_1} + \frac{(\Sigma x_2)^2}{n_2} - \frac{(\Sigma x_T)^2}{n_T} = \frac{789.6}{12} + \frac{615.0}{12} - \frac{2798.4}{24} = 0.454$$

Between groups sum of squares

$$SS_G = \frac{(\Sigma x_{a1})^2}{n_{a1}} + \frac{(\Sigma x_{a2})^2}{n_{a2}} + \frac{(\Sigma x_{b1})^2}{n_{b1}} + \frac{(\Sigma x_{b2})^2}{n_{b2}} + \frac{(\Sigma x_{c1})^2}{n_{c1}} + \frac{(\Sigma x_{c2})^2}{n_{c2}}$$

$$+ \frac{(\Sigma x_{d1})^2}{n_{d1}} + \frac{(\Sigma x_{d2})^2}{n_{d2}} - \frac{(\Sigma x_T)^2}{n_T} = \frac{17.64}{3} + \frac{21.16}{3} + \frac{112.36}{3}$$

$$+ \frac{46.24}{3} + \frac{62.41}{3} + \frac{44.89}{3} + \frac{29.16}{3} + \frac{44.89}{3} - \frac{2798.4}{24} = 9.650$$

Interaction sum of squares

$$SS_I = SS_G - SS_A - SS_B = 9.650 - 6.695 - 0.454 = 2.501$$

Within groups (or residual) sum of squares

$$SS_R = SS_T - SS_A - SS_B - SS_I = 11.710 - 6.695 - 0.454 - 2.501$$
$$= 2.060$$

2. Section 8.4 explains how to set out an ANOVA table of these results. The calculations for degrees of freedom, mean squares, and the F-values are given in the notes below the table. For factor A we get $F_{3,16} = 17.332$. The F-value takes its degrees of freedom from the df for 'island' and the df for 'within groups', in that order. For factor B we get $F_{1,16} = 3.524$. The F-value takes its degrees of freedom from the df for 'beach type' and the df for 'within groups', in that order. For the interaction between factors A and B we get $F_{3,16} = 6.476$. The F-value takes its degrees of freedom from the df for 'island $\times$ beach type' and the df for 'within groups', in that order.

3. From F-tables

for $P = 0.05$, $F_{1,16} = 4.49$, $F_{3,16} = 3.24$
for $P = 0.01$, $F_{1,16} = 8.53$, $F_{3,16} = 5.29$
for $P = 0.001$, $F_{1,16} = 16.12$, $F_{3,16} = 9.01$

Although ANOVA is a two-tailed test – it tests for any differences between means, not just differences in a particular direction – it is not necessary to multiply the P-values obtained from F-tables by 2 when using ANOVA. Therefore, in this case

for factor A, P is less than 0.001, a significant result
for factor B, P is greater than 0.05, a non-significant result
for interaction A $\times$ B, P lies between 0.01 and 0.001, a significant result

Using a computer we find

for factor A, $P < 0.001$
for factor B, $P = 0.079$
for interaction A $\times$ B, $P = 0.004$

Correlation (page 131)

It does not matter which of the variables we call x and which we call y.

x	6.3	4.2	10.3	7.1	8.4	9.5	5.0	3.5	6.4
y	9.4	8.6	6.3	10.9	5.5	8.7	9.8	11.6	7.7
xy	59.22	36.12	64.89	77.39	46.20	82.65	49.00	40.60	49.28

$\Sigma x = 60.7$ $(\Sigma x)^2 = 3684.49$ $\Sigma x^2 = 452.85$

$\Sigma y = 78.5$ $(\Sigma y)^2 = 6162.25$ $\Sigma y^2 = 716.65$

$\Sigma xy = 505.35$

$n = 9$, the number of pairs of measurements

1. Calculate the correlation coefficient, r, with df $= n - 2$

$$r_{n-2} = \frac{\Sigma xy - \dfrac{\Sigma x \Sigma y}{n}}{\sqrt{\left(\Sigma x^2 - \dfrac{(\Sigma x)^2}{n}\right)\left(\Sigma y^2 - \dfrac{(\Sigma y)^2}{n}\right)}}$$

$$= \frac{505.35 - \dfrac{60.7 \times 78.5}{9}}{\sqrt{\left(452.85 - \dfrac{3684.49}{9}\right)\left(716.65 - \dfrac{6162.25}{9}\right)}} = -0.646$$

2. From r-tables

     for $P = 0.05$, $r_7 = 0.666$

     for $P = 0.01$, $r_7 = 0.798$

Note that in some tables of r-values you need to look up r for df $= n - 2$ and in others you need to look up r for a sample size of n, depending on how the tables have been produced. You can use the values here to check you are using the tables correctly. To find the significance, ignore the sign of the r-value, i.e. use $r = 0.646$. Therefore, in this case P is greater than 0.05, a non-significant result. Using a computer we find that $P = 0.060$.

Simple linear regression (page 135)

Note that the variable we call y should be the one that is responding to changes in the other. In regression we will usually be choosing the values of the variable x to be spaced out over the range of interest.

x	0	0.2	0.4	0.6	0.8	1.0	1.2	1.4	1.6	1.8
y	16.6	17.4	18.3	18.2	19.6	20.2	21.7	22.0	22.7	23.2
xy	0	3.48	7.32	10.92	15.68	20.20	26.04	30.80	36.32	41.76

$\Sigma x = 9.0$ $(\Sigma x)^2 = 81.0$ $\Sigma x^2 = 11.4$ $\bar{x} = 0.90$

$\Sigma y = 199.9$ $(\Sigma y)^2 = 39\,960.01$ $\Sigma y^2 = 4045.07$ $\bar{y} = 20.0$

$\Sigma xy = 192.52$

$n = 10$, the number of pairs of measurements

1. Calculate the equation of the regression line

$$\beta = \frac{n\Sigma xy - \Sigma x \Sigma y}{n\Sigma x^2 - (\Sigma x)^2} = \frac{(10 \times 192.52) - (9.0 \times 199.9)}{(10 \times 11.4) - 81.0} = 3.82$$

$$\alpha = \bar{y} - \beta\bar{x} = 20.0 - (3.82 \times 0.90) = 16.6$$

The equation of the line of best fit is $y = \alpha + \beta x$, so we have

$$y = 16.6 + 3.82x$$

2. Calculate sums of squares.

Total sum of squares

$$SS_T = \Sigma y^2 - \frac{(\Sigma y)^2}{n} = 4045.07 - \frac{39\,960.01}{10} = 49.069$$

Regression sum of squares

$$SS_{reg} = \frac{\left(\Sigma xy - \dfrac{\Sigma x \Sigma y}{n}\right)^2}{\Sigma x^2 - \dfrac{(\Sigma x)^2}{n}} = \frac{\left(192.52 - \dfrac{9.0 \times 199.9}{10}\right)^2}{11.4 - \dfrac{81.0}{10}} = 48.185$$

Residual sum of squares

$$SS_{res} = SS_T - SS_{reg} = 49.069 - 48.185 = 0.884$$

3. Section 9.4 shows how to set out an ANOVA table of these results. For the regression, df $= 1$; this is always true for simple linear regression. For the total, df $= n - 1$; and for the residual, df $= n - 2$. The calculation of mean squares and the F-value is the same as for one-way ANOVA (see the ANOVA table in Section 8.3). We get $F_{1,8} = 436.3$.

4. From F-tables

for $P = 0.05$, $F_{1,8} = 5.32$
for $P = 0.01$, $F_{1,8} = 11.26$
for $P = 0.001$, $F_{1,8} = 25.41$

Although regression is a two-sided test – it tests for either a positive or negative association between the variables – it is not necessary to multiply the P-values obtained from F-tables by 2 for this test. Therefore, in this case P is less than 0.001, a significant result. Using a computer we also find that $P < 0.001$.

5. Calculate the coefficient of determination, r^2.

$$r^2 = \frac{SS_{reg}}{SS_T} = \frac{48.185}{49.069} = 0.982$$

Mann–Whitney U-test (page 189)

Sample 1	1.6, 1.6, 1.2, 2.4, 2.8, 1.4	$n_1 = 6$
Sample 2	2.1, 2.9, >3.2, 2.4, 2.8, 2.0, 2.5	$n_2 = 7$

1. Rank the values, noting which come from which sample.
   (sample 1 values in bold)

Value	**1.2**	**1.4**	**1.6**	**1.6**	2.0	2.1	**2.4**	2.4	2.5	**2.8**	2.8	2.9	>3.2
Rank	**1**	**2**	**3.5**	**3.5**	5	6	**7.5**	7.5	9	**10.5**	10.5	12	13

Sum of ranks for sample 1, $R_1 = 28$
Sum of ranks for sample 2, $R_2 = 63$

2. Some statistical tables allow you to look up these 'rank sums' directly. These are more likely to be known as tables for the Wilcoxon rank sum test, which gives identical results.

3. Calculate U. There are various ways to do this; here is one of them:

$$U_1 = R_1 - \frac{n_1(n_1 + 1)}{2} = 28 - \frac{6 \times 7}{2} = 7$$

$$U_2 = R_2 - \frac{n_2(n_2 + 1)}{2} = 63 - \frac{7 \times 8}{2} = 35$$

U is the smaller of these two values:

$$U = \min(U_1, U_2) = \min(7, 35) = 7$$

4. From Mann–Whitney U-tables

for $P = 0.05$ with $n_S = 6$, $n_L = 7$ we have $U = 6$
for $P = 0.01$ with $n_S = 6$, $n_L = 7$ we have $U = 3$

where n_S is the smaller of the two sample sizes, and n_L is the larger of the two sample sizes. Unlike most statistical tests, the result is significant if the U-value we calculate is *smaller* than the value of U given in the tables. In this case P is greater than 0.05, a non-significant result. Using a computer we find that $P = 0.054$.

Kruskal–Wallis test (page 195)

Sample 1	1.6, 1.6, 1.2, 2.4, 2.8, 1.4	$n_1 = 6$
Sample 2	2.1, 2.9, >3.2, 2.4, 2.8, 2.0, 2.5	$n_2 = 7$
Sample 3	1.1, 1.0, 0.9, 1.5, 1.2	$n_3 = 5$

1. Rank the values, noting which come from which sample.
   (sample 1 in bold, sample 2 in normal font, sample 3 in italics)

Value	0.9	1.0	1.1	**1.2**	1.2	**1.4**	1.5	**1.6**	**1.6**
Rank	1	2	3	**4.5**	4.5	**6**	7	**8.5**	**8.5**

Value	2.0	2.1	**2.4**	2.4	2.5	**2.8**	2.8	2.9	> 3.2
Rank	10	11	**12.5**	12.5	14	**15.5**	15.5	17	18

Sum of ranks for sample 1, $R_1 = 55.5$
Sum of ranks for sample 2, $R_2 = 98$
Sum of ranks for sample 3, $R_3 = 17.5$

$$\frac{R_1^2}{n_1} = 513.38 \quad \frac{R_2^2}{n_2} = 1372.00 \quad \frac{R_3^2}{n_3} = 61.25 \quad \sum\left(\frac{R^2}{n}\right) = 1946.63$$

$N = n_1 + n_2 + n_3 = 6 + 7 + 5 = 18$, the total number of values in the dataset

2. Calculate the test statistic, H; other books may use a different symbol.

$$H = \frac{12}{N(N+1)}\sum\left(\frac{R^2}{n}\right) - 3(N+1)$$

$$= \frac{12}{18 \times 19} \times 1946.63 - (3 \times 19) = 11.30$$

3. For very small samples (i.e. when $N \leq 16$), special tables for the Kruskal–Wallis test should be used. For larger samples, H can be looked up in χ^2 tables with degrees of freedom = number of samples − 1, so here df = 3 − 1 = 2.

4. From χ^2 tables

for $P = 0.05$, $\chi_2^2 = 5.99$
for $P = 0.01$, $\chi_2^2 = 9.21$
for $P = 0.001$, $\chi_2^2 = 13.82$

Therefore, in this case P lies between 0.01 and 0.001, a significant result. Using a computer we find that $P = 0.004$.

Friedman's test (page 198)

In the example there is one factor, position on slope, with three levels: bottom (level 1), middle (level 2) and top (level 3). The term 'levels' applies whatever the factor, e.g. factors such as colour, pH and country would all be divided into levels. The term 'treatments' may also be used instead of 'levels'. Measurements were taken on several days. We are not interested in differences between days but we want to allow for them. Days are therefore the blocks in the design.

	Factor		
	Level 1	Level 2	Level 3
Block 1	6	5	8
Block 2	3	10	8
Block 3	4	15	26
Block 4	6	26	100

1. Rank the values within each block.

	Factor		
	Level 1	Level 2	Level 3
Block 1	2	1	3
Block 2	1	3	2
Block 3	1	2	3
Block 4	1	2	3

2. Sum of ranks for level 1, $R_1 = 5$
   Sum of ranks for level 2, $R_2 = 8$
   Sum of ranks for level 3, $R_3 = 11$

$R_1^2 = 25 \quad R_2^2 = 64 \quad R_3^2 = 121 \quad \Sigma R^2 = 210$

$l = 3$, number of levels of the factor
        (or the number of treatments in an experiment)
$b = 4$, number of blocks

3. Calculate the test statistic, S; other books may use a different symbol.

$$S = \frac{12}{bl(l+1)}\Sigma R^2 - 3b(l+1) = \frac{12}{4 \times 3 \times 4} \times 210 - (3 \times 4 \times 4) = 4.5$$

4. For very small samples, (i.e. when $b \times l \le 25$) special tables for Friedman's test should be used. For larger samples, the value S can be looked up in χ^2 tables, with degrees of freedom = number of levels of the factor -1, i.e. in this case df $= 3 - 1 = 2$. As we have very small samples, in this case, tables for the Friedman test should be used.

5. From Friedman test tables

   for $P = 0.05$ with $l = 3$, $b = 4$ we have $S = 6.5$
   for $P = 0.01$ with $l = 3$, $b = 4$ we have $S = 8.0$

Therefore, in this case P is greater than 0.05, a non-significant result. Using a computer we find that $P = 0.105$.

Spearman's rank correlation (page 200)

In this example there are two variables. Conventionally we would call road size score x and Pb concentration y because we believe road size is controlling Pb concentration, not vice versa. However, for the calculation of Spearman's rank correlation coefficient, it does not matter which we call x.

x	1	3	2	3	1	4	4	2
y	12	16	12	90	3	27	44	4

1. Rank the data *within* each variable then calculate pairwise differences $d = (x_r - y_r)$.

x_r	1.5	5.5	3.5	5.5	1.5	7.5	7.5	3.5
y_r	3.5	5	3.5	8	1	6	7	2
d	-2.0	0.5	0	-2.5	0.5	1.5	0.5	1.5
d^2	4.0	0.25	0	6.25	0.25	2.25	0.25	2.25

$\Sigma d^2 = 15.5$

$n = 8$, the number of pairs of observations

2. Calculate Spearman's rank correlation coefficient, r_S.

$$r_S = 1 - \frac{6}{n^3 - n}\Sigma d^2 = 1 - \frac{6}{512 - 8} \times 15.5 = 0.82$$

3. From tables of Spearman's rank correlation coefficient

for $P = 0.05$ with $n = 8$ we have $r_S = 0.738$
for $P = 0.01$ with $n = 8$ we have $r_S = 0.881$

Therefore, in this case P lies between 0.05 and 0.01, a significant result. Using a computer we find that $P = 0.025$.

Appendix B

Concentration Data for Chapters 14 and 15

The following table gives the concentrations in micrograms per gram ($\mu g\,g^{-1}$) of Subst1 to Subst11 at each of the 20 factory sites.

Factory	Subst1	Subst2	Subst3	Subst4	Subst5	Subst6	Subst7	Subst8	Subst9	Subst10	Subst11
1	3.3	1.95	84	0.0388	8.9	3.5	625	0.7	112	3.6	1.0
2	5.2	3.00	159	0.0325	8.1	2.6	427	2.4	86	2.8	2.0
3	3.8	1.60	90	0.0400	9.5	3.4	741	3.4	152	5.0	51.5
4	5.6	2.85	156	0.0288	7.8	2.6	485	9.2	244	8.2	1.7
5	5.3	2.60	180	0.0313	8.4	2.4	472	11.0	224	7.3	51.6
6	5.5	2.65	171	0.0325	8.9	2.7	565	2.3	136	4.4	50.0
7	4.0	1.65	81	0.0288	8.6	3.7	677	0.7	65	2.4	2.2
8	3.5	1.60	99	0.0375	8.6	3.1	645	6.8	230	7.8	50.8
9	5.6	2.90	153	0.0325	7.9	2.2	510	2.4	116	3.6	52.4
10	3.0	1.55	114	0.0400	9.4	2.6	633	6.2	203	6.3	1.2
11	5.2	2.80	171	0.0300	8.2	2.8	550	2.6	133	4.3	51.0
12	3.5	1.95	75	0.0325	8.7	3.3	563	6.0	194	6.1	2.0
13	3.5	1.85	84	0.0363	9.2	3.2	599	9.6	274	8.2	3.9
14	5.6	2.85	156	0.0338	8.1	2.6	429	10.0	250	8.2	2.7
15	3.1	1.65	108	0.0388	8.5	3.1	672	8.8	217	7.4	51.0
16	4.0	1.55	87	0.0313	9.2	3.2	664	4.4	115	4.2	3.2
17	5.4	2.70	171	0.0263	8.8	1.8	502	0.9	43	1.9	2.5
18	5.5	2.60	174	0.0238	8.2	1.8	486	10.5	278	8.4	1.0
19	5.5	2.60	174	0.0350	8.1	2.8	607	1.1	60	2.6	1.4
20	3.5	1.50	105	0.0400	9.6	3.1	793	13.8	300	9.2	1.6

Appendix C
Using Computer Packages

This appendix contains notes about how to carry out most of the analyses described in this book using Excel and Minitab, although other packages are referred to briefly at the end. Both of these packages contain extensive instructions in their help files, accessible by opening the **Help** menu at the top of the screen, or by clicking on buttons labelled '?' or 'Help' where these appear while using the programs. These will give you details about how to use the functions and commands described briefly below. About half the tests can be carried out using Excel, and all except nonlinear regression, the Kolmogorov–Smirnov test and Spearman's rank correlation test can be carried out using Minitab. The notes in this appendix relate specifically to Microsoft Excel 2000 and Minitab for Windows release 13.20 but most will apply to earlier versions, and probably to later versions as well.

C.1 Excel

Excel includes some useful statistical functions which can be entered into a cell as part of a formula. A list of these can be accessed by clicking the function button labelled f_x on the **Standard** toolbar. Here are some useful ones:

- `AVERAGE` calculates the mean.
- `CHIDIST` gives P-values for given values of χ^2 and df. It can be used to look up the result of a χ^2 test or the Kruskal–Wallis and Friedman tests for large samples.
- `CHITEST` carries out a χ^2 test for association given both the observed and expected values. The function gives a P-value directly.
- `FDIST` gives P-values for given values of F and df. It can be used to look up the result of ANOVA or an F-test; for the F-test, multiply the P-value by 2.

- STDEV calculates the standard deviation.

- TDIST gives P-values for given values of t and df. Unless you have good reason to do otherwise, specify two tails, i.e. that you require the two-tailed P-value. It can be used to look up the result of a t-test.

- TINV gives t-values for a given probability and df. It can be used to obtain the t-value required to calculate a least significant difference or a confidence interval.

- VAR calculates the variance.

Most of the statistical tests are accessed by opening the **Tools** menu and selecting **Data Analysis**. If this does not appear on the **Tools** menu, first select **Add Ins**, also from the **Tools** menu. Then select **Analysis ToolPak** and click **OK**. Next time you open the **Tools** menu, **Data Analysis** should appear. Using a stand-alone PC it should only be necessary to do this once. Using a networked PC it may be necessary to do this at the start of each session.

Most of the tests in Excel require the data for each sample to be in a separate column (or row). For a t-test the data might be arranged like this:

Sample 1	Sample 2
14	10
6	16
10	12
12	14
9	18
11	15

Many of the tests allow you to select the actual numbers only, or the numbers as well as the names at the head of the columns, e.g. 'Sample 1'. If you choose the latter, you also need to tick the box *Labels in first row* or *Labels* in the dialog box for the test. The output from the test will then include your variables' names. After selecting **Data Analysis** from the **Tools** menu, the following commands should be available.

Experimental design

- **Random Number Generation** can be used to produce sets of random numbers that you can then use for setting out a randomized design or selecting a sample. Set the distribution as Uniform.

Descriptive statistics

- **Descriptive Statistics** gives a range of useful summary statistics for a sample, including the standard deviation, variance, mean, median, standard error and 95% confidence interval; actually the 95% CI is the sample mean $\pm$ the value given here by Excel.

- **Histogram** plots a frequency distribution. By default the function chooses a convenient number of bars to include on the graph. This can be overridden so that the bars represent specific ranges of values by specifying the 'bins' to use; see the program's help file for details.

t-tests

There is no command for a one-sample *t*-test but the test can be carried out using the paired *t*-test. Suppose you have a sample with eight values and you want to compare the population mean with a fixed value of, say, 3.5. Enter the value 3.5 eight times in an adjacent column and use this as the second sample in the paired *t*-test. All of these tests produce both one-tailed and two-tailed *P*-values. In general, you should use the two-tailed *P*-value.

- *t*-**Test: Paired Two Sample for Means** carries out a paired *t*-test.

- *t*-**Test: Two-Sample Assuming Equal Variances** carries out an unpaired *t*-test using the formula in Appendix A.

- *t*-**Test: Two-Sample Assuming Unequal Variances** carries out an unpaired *t*-test but makes an adjustment to the *P*-value if the samples do not have equal variance. The name is inaccurate in that it should say 'not assuming equal variances'. It can be used whether or not the populations have equal variance. If the samples have the same variance, it gives identical results to the test above.

F-test

F-Test Two-Sample for Variances carries out a one-tailed *F*-test. Unless you have good reason to want to carry out a one-tailed test, you should multiply the *P*-value by 2 to get the result for a two-tailed test.

ANOVA

- **Anova: Single Factor** carries out a one-way ANOVA. Samples should be arranged in adjacent columns. Sample sizes do not need to be equal for the calculations to be carried out, though they should be similar for the test result to be meaningful.

- **Anova: Two-Factor With Replication** carries out a two-way ANOVA. Data must be arranged in a particular format. For an experiment in which one factor has three levels (pH 6, pH, 7 and pH 8), the other has two levels (hilltop and valley), and there are three replicates, you would use the following layout:

	pH 6	pH 7	pH 8
Hilltop	5	6	7
	4	7	7
	4	8	9
Valley	3	6	8
	5	6	7
	5	8	10

The input range would then be a rectangle of cells four columns wide and seven rows deep (i.e. it includes the data plus one row of headings above and one column of headings to the left). The number of 'rows per sample' to enter is the number of replicates, in this case three. The sample sizes must be equal and there must be no missing values. The output does not include the residuals. To calculate the residuals to check for a Normal distribution, subtract the mean for each treatment combination from the values in that treatment combination (e.g. for hilltop, pH6, calculate the mean $= 4.333$, then subtract it from each of the values 5, 4 and 4 in turn). We get 0.667, -0.333 and -0.333. Carry this out for each treatment combination to get the 18 residuals then check that these have an approximately Normal distribution.

- **Anova: Two-Factor Without Replication** can be used to analyse the data from a randomized complete blocks design with one factor of interest (e.g. watering treatment), one blocking factor (e.g. time of day) and one individual from each treatment in each block. Data should be entered in the same format as for two-way ANOVA. The input range should include the row and column headings and you should tick the *Labels* checkbox in the test's dialog box. If you have used more than one individual from each treatment in each block, use two-way ANOVA with replication instead (see above).

Correlation and regression

- **Correlation** calculates Pearson's product moment correlation coefficient. If you have measured several variables, this command can also be used to produce a matrix of correlations between each of them in pairs. *P*-values are not given.

- **Regression** carries out a regression analysis; *y*-values (i.e. values of the dependent variable) should be entered in one column on the spreadsheet, and *x*-values (i.e. values of the independent variable) in another. The command can be used to carry out simple linear regression or multiple linear regression. If there are several *x*-variables (i.e. several independent variables) these should be entered in adjacent columns and all of these should be selected together for *Input X Range*. Polynomial regression can be carried out by first calculating columns of values of x^2, x^3, etc,. alongside the column containing the *x*-values. These columns are then all selected together as *Input X Range*, as for multiple linear regression.

  Ticking the box labelled *Residuals* will save the residuals from the analysis into a column on the spreadsheet. These can be used to check for a Normal distribution. The output includes r^2 (*R Square*). The middle table is labelled *ANOVA* and the value *Significance F* in this table is the *P*-value for the significance of the regression. The line of best fit is given by the *Coefficients* in the third table of the output.

Non-parametric tests

None of the non-parametric tests covered in Chapter 13 are available as functions or commands in Excel.

Multivariate methods

None of the multivariate methods covered in this book are available as functions or commands in Excel.

C.2   Minitab

The tests are accessed through the **Stat** menu, which lists tests in groups such as **Nonparametrics**, within which the individual tests can be found, e.g. **Mann–Whitney**. In the notes below this is designated **Stat ▶ Nonparametrics ▶ Mann–Whitney**. Many of the tests have the option to display graphical summaries of the data at the same time. These may be useful for checking that the assumptions of the test are met.

Two formats for data are used within Minitab. A few tests require you to enter the data for different samples in adjacent columns, as described above for Excel; Minitab calls this an 'unstacked' format. Most, however, require the data for the different samples to be in the same column, a 'stacked' format. Further columns are then used to indicate which group an individual value comes from; so for a *t*-test, instead of entering the data in two columns as shown above, we could enter it like this. The names you give to the columns are up to you; I have called them data and population.

Data	Population
14	1
6	1
10	1
12	1
9	1
11	1
10	2
16	2
12	2
14	2
18	2
15	2

The data for the two-way ANOVA example above would be entered as follows:

Data	pH	Position
5	6	hilltop
4	6	hilltop
4	6	hilltop
3	6	valley
5	6	valley
5	6	valley
6	7	hilltop
7	7	hilltop
8	7	hilltop
6	7	valley
6	7	valley
8	7	valley
7	8	hilltop
7	8	hilltop
9	8	hilltop
8	8	valley
7	8	valley
10	8	valley

If you have more treatment factors or one or more blocking factors, these can be entered in additional columns. For multivariate data, each variable should be entered in a separate column, like those in Appendix B.

Miscellaneous commands

- **Calc ▶ Calculator** is used to perform the same calculation on each number in a column of numbers. This can be used for transforming data, e.g. entering the formula `LOGT(c1+1)` would carry out a log transformation of the values in column 1.

- **Manip ▶ Rank** converts a set of values into their equivalent ranks.

- **Stat ▶ Control Charts ▶ Box-Cox Transformation** selects the best value of λ to achieve a Normal distribution and carries out the transformation. The Minitab formula looks slightly different to the formula in Section 6.3 but it achieves the same thing.

- **Stat ▶ Power and Sample Size** can be used to help estimate the best sample size to use, or the power of a test for a given sample size, for t-tests and one-way ANOVA, and for multiway ANOVA when each factor has only two levels.

- **Stat ▶ ANOVA ▶ Test for Equal Variances** carries out Bartlett's test and Levene's test for equal variances. This could be used to help check whether the assumption of equal variance in an ANOVA is contravened by your data.

Experimental design

- **Calc ▶ Random Data ▶ Integer** can be used to produce sets of random numbers which you can use for setting out a randomized design or selecting a sample.

Descriptive statistics

- **Stat ▶ Basic Statistics ▶ Display Descriptive Statistics** calculates various summary statistics, including sample means, medians, standard deviations, and standard errors. There is also the option to plot histograms of the samples and carry out tests for Normality using the option **Graphs ▶ Graphical Summary**.

t-tests

- **Stat ▶ Basic Statistics ▶ 1-Sample t** carries out a one-sample *t*-test.
- **Stat ▶ Basic Statistics ▶ 2-Sample t** carries out an unpaired *t*-test.
- **Stat ▶ Basic Statistics ▶ Paired t** carries out a paired *t*-test.

By default these tests are two-tailed. One-tailed tests can be carried out by selecting **Options ▶ Alternative**.

F-test

- **Stat ▶ Basic Statistics ▶ 2 Variances** carries out an *F*-test. The *P*-value given is for a two-sided test so it does not need to be multiplied by 2.

ANOVA

- **Stat ▶ ANOVA ▶ One-way** carries out one-way ANOVA. This command will also carry out multiple comparisons of means to determine which means differ. There is an option to store the residuals for checking that the assumptions of the test are met. Graphs of the residuals can also be produced directly using the option **Graphs**. For *Response* enter the column containing your measurements.

- **Stat ▶ ANOVA ▶ Two-way** carries out two-way ANOVA. As for one-way ANOVA, there are options to store and display graphs of the residuals. No missing values are allowed. For *Response* enter the column containing your measurements.

- **Stat ▶ ANOVA ▶ General Linear Model** can be used for all types of ANOVA. Unequal sample sizes and missing values are allowed. The price is that it is slightly more complicated to run the command. For *Responses* enter the column containing your measurements, usually c1. In the box labelled *Model*, enter c2 (for a one-way ANOVA), c2!c3 (for a two-way ANOVA), c2!c3!c4 (for a three way ANOVA), etc., where c2, c3 and c4 are columns containing the different factors in the experiment. For a two-way ANOVA with blocks, enter c2!c3+c4, where c4 contains the blocks. To carry out pairwise comparisons of the means as well, select **Comparisons** and in the box labelled *Terms* enter the same expression that you have entered for *Model*. You can also use the **Graphs** option to obtain plots of the residuals to check that the assumptions of the test are met by your data.

Correlation and regression

- **Stat ▶ Basic Statistics ▶ Correlation** can be used to calculate Pearson's product moment correlation coefficient between pairs of variables.

- **Stat ▶ Regression ▶ Regression** carries out a regression analysis. *Response* is the column containing the dependent or *y*-variable. *Predictors* are one or more columns containing the independent or *x*-variable(s). The **Graphs** option can be used to plot graphs of the residuals to check that the assumptions of the test are met by the data. The output includes the equation of the line of best fit, r^2, and the adjusted r^2. The overall significance of the regression is the *P*-value given in the table headed *Analysis of Variance*.

- **Stat ▶ Regression ▶ Fitted Line Plot** draws a graph of the line of best fit for a simple linear regression. Confidence limits for the line can also be shown on the graph by selecting **Options** and *Display confidence bands*.

- **Stat ▶ Regression ▶ Best Subsets** selects the best of several possible *x*-variables to include in a regression. The output gives r^2 and the adjusted r^2 for the best one, the best two, the best three, etc. *x*-variables to include. By default the best two options are displayed for each number of *x*- variables to include. This can be changed by selecting **Options** and *Models of each size to print*. The column *Response* is the column containing the *y* or dependent variable. *Free predictors* are the columns containing the *x* or independent variables that you want to try in the regression.

Chi-square test

- **Stat ▶ Tables ▶ Chi-square Test** carries out a chi-square test for association. The observed values should be entered into a series of columns in the spreadsheet like a table of rows and columns. The test calculates the expected values and then the *P*-value. This command could also be used to compare proportions.

Non-parametric tests

- **Stat ▶ Nonparametrics ▶ Mann-Whitney** carries out a Mann–Whitney *U*-test, also known as a Wilcoxon rank sum test. The test requires the data for the two samples being compared to be in separate columns. The *P*-value is given as 'the test is significant at . . .' with and without adjustment for tied values. In general use the adjusted result.

- **Stat ▶ Nonparametrics ▶ Kruskal-Wallis** carries out a Kruskal–Wallis test.

Response is the column containing your actual measurements or observations. *Factor* is a column containing the treatments from which each individual came. The test statistic, H, is calculated and looked up by the program as a chi-square value to find the *P*-value. For very small samples – i.e. when the total number of measurements made ≤ 16 – this can give an inaccurate result. It would be better in these cases to take the calculated value of H and look this up in Kruskal–Wallis tables (e.g. Neave 1978).

- **Stat ▶ Nonparametrics ▶ Friedman** carries out Friedman's test. *Response* is the column containing your actual measurements or observations. *Treatment* is a column containing the treatment from which each individual came. *Blocks* is a column containing the block in which each individual was located. The test statistic, S, is calculated and looked up by the program as a chi-square value to find the *P*-value. For very small samples, – i.e. when the total number of measurements made ≤ 25 – this can give an inaccurate result. It would be better in these cases to take the calculated value of S and look this up in Friedman tables (e.g. Neave 1978).

- **Stat ▶ Nonparametrics ▶ 1-Sample Sign**: there is no command for a two-sample sign test as such, but this can be carried out using the one-sample sign test command. Arrange the data for the two samples in two columns (e.g. c1 and c2). Then calculate a column of differences (c1–c2) using **Calc ▶ Calculator** (see Miscellaneous commands) and store these in c3. In the one-sample sign test dialog box, enter c3 as the *Variables*, select *Test median* as 0.0, and *Alternative* as not equal. The *P*-value given by the test is then the same as if we had carried out a two-sample sign test.

Minitab has no command for Spearman's rank correlation or for the Kolmogorov–Smirnov test. It is possible to calculate the value of Spearman's rank correlation coefficient, r_s, by first converting the values to ranks using **Manip ▶ Rank** (see Miscellaneous commands), then calculating the correlation coefficient for these using **Stat ▶ Basic Statistics ▶ Correlation**. However, the *P*-value given will not be correct; r_s must be looked up in Spearman's rank correlation tables to obtain the correct *P*-value (e.g. Neave 1978).

Multivariate methods

- **Stat ▶ ANOVA ▶ General MANOVA** carries out multivariate analysis of variance (MANOVA). *Responses* and *Model* are as for **Stat ▶ ANOVA ▶ General Linear Model** (see above), except that you will enter more than one column for *Responses*, i.e. all of the columns containing your measurements.

- **Stat ▶ Multivariate ▶ Principal Components** carries out principal component analysis. For *Variables* enter the columns containing the different types of measurements you want to include in the analysis. For most purposes, six is a reasonable number of components to compute. In the **Graphs** option you can select to output a scree plot and a scatter graph of the scores on the first two principal components. The output gives eigenvalues and cumulative proportions of variance explained by successive principal components. The table below this contains the eigenvectors (also called coefficients or loadings) used to calculate the scores. These can also be stored on the spreadsheet by selecting **Storage** and entering the required columns in which to store them in the box labelled *Coefficients*. The scores themselves can also be stored on the spreadsheet by selecting **Storage** and entering the required columns in which to store them in the box labelled *Scores*.

- **Stat ▶ Multivariate ▶ Cluster Observations** carries out cluster analysis for the individuals in the sample. The columns containing the different types of measurement should be entered under *Variables or distance matrix*. For most purposes you should also select *Standardize variables* and *Show dendrogram*.

- **Stat ▶ Multivariate ▶ Cluster Variables**: the columns containing the different types of measurement should be entered under *Variables or distance matrix*. For most purposes you should select *Correlation* under *Distance Measure* and choose *Show dendrogram*.

C.3   Other packages

SPSS for windows is another popular statistical program and in many ways is similar to Minitab for Windows to use.

Genstat and SAS are often favoured by those who prefer to use command lines – short instructions typed into a file or entered directly from the keyboard – rather than the Windows menu system. One advantage of this approach is the ability to include instructions telling the program how the data is set out in the original file, and how the output should be formatted. This can be useful if, for instance, you are reading in values from a data logger or outputting values to a graphics program. These programs are very versatile but generally considered more difficult to learn.

Many of the statistical tests described in this book can be carried out using resources available free on the World Wide Web. Various web pages exist in which you can enter your data, click a button to submit it, and receive the result of the test directly. Since web pages come and go, I will not refer to any specifically here, but they can be tracked down with a modest amount of effort using any of the major search engines.

Appendix D
Choosing a Test: Decision Table

The following table summarizes some of the main requirements of the tests and methods described in this book, and situations in which they may be used. The table is intended to direct you towards an appropriate technique but you should consult the relevant section to confirm its suitability. Most of the tests have other requirements as well as those listed here – such as the requirement for samples to be made up of independent and randomly selected individuals – and these are covered as part of the descriptions of the tests. If no suitable technique can be found from this table, you might find ideas in the further reading suggestions at the end of some of the chapters. However, you should not assume that there are tests for all eventualities. To avoid disappointment, it is far better to select a test before collecting any data.

Table D.1 Decision table for choosing a test

Name of test or method	Test for	Number of variables
Unpaired *t*-test	Difference between typical members of populations	1
Paired *t*-test	Difference between typical members of a population	1
One-sample *t*-test	Difference between typical member of a population and a fixed value	1
F-test	Difference in spread of values in populations	1
One-way ANOVA	Difference between typical members of populations	1
Multiway ANOVA	Difference between typical members of populations	1
Correlation	Association between variables	2
Simple linear regression	Association between variables	2
Multiple regression	Association between variables	3+
Comparing lines	Differences in association between variables	2
MANOVA	Difference between typical members of populations	2+
Chi-square	Goodness of fit	1
Chi-square	Association between factors	1
Chi-square	Difference in proportions	1
Mann–Whitney	Difference between typical members of populations	1
Kolmogorov–Smirnov	Difference between typical members of populations	1
Sign test	Difference between typical members of populations	1
Kruskal–Wallis	Difference between typical members of populations	1
Friedman	Difference between typical members of populations	1
Spearman's rank correlation	Association between variables	2
Principal component analysis	n/a	2+
Cluster analysis	n/a	2+

n/a = not applicable; **2+** = two or more

Typical: here a 'typical' member of a population means the mean or median

Spread: here the term 'spread' means the variance or standard deviation

Number of variables: the number of types of measurements or observations we make on each individual in the samples (e.g. height = 1 variable; height and mass = 2 variables; height, mass and age = 3 variables)

Number of factors	Number of populations	Must individuals in different populations be paired?	Are Normal distributions required?	Must populations have the same shape of distribution?	Section in this book
1	2	no	yes	yes	7.3
1	2	yes	yes	no	7.3
n/a	1	n/a	yes	n/a	7.3
1	2	no	yes	no	7.4
1	2+	no	yes	yes	8.3
2+	4+	no	yes	yes	8.4
n/a	1	n/a	yes	n/a	9.3
n/a	1	n/a	yes	n/a	9.4
n/a	1	n/a	yes	n/a	9.6
1	2	no	yes	yes	9.7
1+	2+	no	yes	yes	10.1
n/a	1	n/a	no	n/a	12.3
2	1	n/a	no	n/a	12.4
1	2+	no	no	no	12.5
1	2	no	no	yes	13.3
1	2	no	no	no	13.4
1	2	yes	no	no	13.5
1	2+	no	no	yes	13.6
1 + blocks	2+	no	no	yes	13.7
n/a	1	n/a	no	n/a	13.8
n/a	1+	n/a	no	n/a	14.1
n/a	1+	n/a	no	n/a	15.1

Number of factors: the number of different ways in which individuals are grouped in the experiment or survey. For example, if we include different varieties of tomato plants, and a number of different watering treatments and fertilizer treatments are applied to each of these, the experiment includes three factors: variety, fertilizer and watering treatment

Number of populations: the number of groups we are comparing or studying. In the case of multiway ANOVA there must be at least four groups present in the experiment, i.e. high and low levels of factor 1, each combined with high and low levels of factor 2

Appendix E
List of Worked Examples

The text includes a number of examples used to illustrate the different techniques. By drawing parallels between these examples and your own data, a suitable technique to use can be identified. Note that the data in the examples in this book are made up to demonstrate the techniques and are not necessarily scientifically accurate results. Calculations for many of these examples are also given in Appendix A.

Unpaired *t*-test (page 95) – Comparing means of two populations

A randomly selected sample of hedgehogs is observed in each of two areas and the number of days each animal spends in hibernation is recorded. An unpaired *t*-test is used to test for a difference in the mean numbers of days spent in hibernation by hedgehogs in the two different areas.

Paired *t*-test (page 97) – Comparing means of two populations using paired samples

SO_2 concentrations are measured at a set of randomly selected locations in one year and then at the *same* locations the following year. A paired *t*-test is used to test whether mean SO_2 concentrations in the air have changed.

One-sample *t*-test (page 100) – Comparing the mean of a population with a fixed value

A sample of plants growing in hydroponic culture are given nitrogen which is 3% enriched with ^{15}N. The ratio of enriched to total nitrogen remaining in the nutrient solution of each plant at the end of the experiment is measured to determine whether it is still 3% (i.e. that plants have not discriminated in favour of or against ^{15}N). A one-sample t-test is used to test whether the mean ^{15}N concentration in plants' nutrient solutions at the end of the experiment is 3%.

F-test (page 103) – Comparing variances of two populations

A randomly selected sample of hedgehogs is observed in each of two areas and the number of days each animal spends in hibernation is recorded. An F-test is used to test for a difference in the variances of numbers of days spent in hibernation by hedgehogs in the two different areas.

One-way ANOVA (page 111) – Comparing means of several populations

Replicated plots of five varieties of millet are grown in a field, set out in a randomized layout. One-way analysis of variance (ANOVA) is used to test for any differences between the mean yields of the different varieties. The example goes on to establish which of the varieties have greater mean yields than the others.

Two-way ANOVA (page 121) – Testing for the effects of two different factors and whether they interact

Densities of litter are recorded on random samples of different types of beach, on four different islands in a small group of islands in the Pacific. Two-way analysis of variance (ANOVA) is used to test for differences in the density of litter on different beach types and on the different islands. The test also tests for an *interaction*, i.e. whether any differences in the densities of litter on different beach types are consistent on the different islands.

Correlation (page 131) – Testing whether there is a straight-line relationship between two variables

The number of hours slept during the day and the number of hours slept the following night is recorded for a random sample of babies. A correlation test is used to test whether there is a linear relationship between the number of hours babies sleep in the day and the number of hours they sleep the following night.

Regression (page 136) – Determining an equation to describe the relationship between two variables

Clutches of flies' eggs are incubated at a series of different saturation deficits (i.e. humidities) and the median time to hatching is recorded in each case. Linear regression is used to test whether there is a straight-line relationship between median hatching time and saturation deficit, and to produce an equation for a line describing this relationship.

Multiple regression (page 143) – Determining an equation to describe how values of one variable can be estimated from two or more other variables

In an extension of the above example, flies' eggs are incubated at a series of different saturation deficits in combination with a series of different light intensities. Multiple regression is used to test whether it is useful to measure both saturation deficit and light intensity to estimate hatching times, and to produce an equation that would allow us to estimate hatching time from saturation deficit and light intensity.

Comparing two regression lines (page 146) – Testing whether two variables show the same linear relationship in two different populations

A random sample of ex-university lecturers who have been admitted to mental institutions is selected. Regression is used to produce equations relating the age at which they were admitted to the institution to the number of hours lecturing they did in their final year of work, and a test is carried out to determine whether this relationship is the same for arts lecturers and science lecturers

Multivariate ANOVA (page 156) – Comparing several populations using multivariate data

Random samples of trees are selected from three populations. On each tree, branch density, bark thickness and height are measured. Multivariate analysis of variance (MANOVA) is used to test whether there are any differences between the populations considering all of these variables together.

Repeated measures (page 166) – Analysing repeated measures data

Measurements of photosynthesis are made on samples of plants growing in four different CO_2 × water treatments. The *same* leaves are measured on a number of occasions. Various approaches to comparing different treatments and times are described.

Chi-square goodness of fit test (page 175) – Testing whether individuals are distributed amongst categories according to some proposed distribution

The numbers of seeds landing in small plots at different distances downwind from a group of trees are compared to the numbers that would be expected if seed dispersal followed an inverse square law, a distribution that has been suggested in this case. A chi-square test is used to test whether the observed counts are consistent with the suggested distribution.

Chi-square test for association (page 178) – Testing whether there is evidence that two particular characteristics or traits in individuals tend to occur together

Moth traps are set up in different habitat types and the numbers of moths of three different species trapped in each habitat are recorded. A chi-square test is used to test whether some species of moth are more or less associated with particular habitat types than the others.

Chi-square test to compare proportions (page 182) – Comparing proportions of individuals that have a particular characteristic, in two (or more) populations

The numbers of sheep which have still-born lambs in an area is recorded in two different years. A chi-square test is used to test whether the proportion of sheep having a still-born lamb differs between years.

Mann-Whitney *U*-test (page 189) – Comparing medians of two populations

Soil penetration resistance is measured at a randomly selected sample of locations in two fields. Some values are off-scale so it is not possible to use a *t*-test. A Mann-Whitney *U*-test is used to test for a difference in median soil penetration resistance between the two fields.

Kolmogorov-Smirnov test (page 191) – Comparing distributions

The time to germination is recorded for each seed in samples of seeds taken from two different batches. A two-sample Kolmogorov-Smirnov test is used to test whether the distributions of times to germination in the two batches differ.

Sign test (page 194) – Comparing medians of two populations using paired samples

Numbers of aphids are counted on a randomly selected sample of leaves of plants growing in a greenhouse. After fumigating the greenhouse the numbers of aphids are counted on the *same* leaves. The differences between before and after have a very non-Normal distribution so it is not possible to use a paired t-test. A two-sample sign test is used to test for a difference in median numbers of aphids per leaf before and after fumigation.

Kruskal-Wallis test (page 196) – Comparing medians of several populations

Using the same example as the Mann-Whitney U-test (see above), a Kruskal-Wallis test is used to test for any differences between the median soil penetration resistances in three different fields.

Friedman's test (page 198) – Comparing medians of several populations in a randomized complete blocks design

Soil matric suction (a measure of soil dryness) is measured at a number of randomly selected points at each of three different heights on a slope. One set of measurements is made on each of several days. We are not interested in the differences between days themselves but we want to allow for them. Days can therefore be treated as 'blocks'. The distributions of measurements are very non-Normal so it is not possible to use ANOVA. Friedman's test is used to test for any differences in median soil matric suction at different heights, allowing for the fact that the measurements are made on different days.

Spearman's rank correlation (page 200) – Testing whether there is a relationship (not necessarily a straight line) between two types of measurement

Atmospheric lead concentration is measured at the sides of different categories of road. Road category is recorded as a score from 1 (minor) to 4 (motorway). As this is not an actual measurement, ordinary correlation cannot be used. Spearman's rank correlation is used to test for a relationship between road category and lead concentration.

Principal component analysis (page 207) – Identifying the most variable characteristics in populations and groupings of individuals

Soil samples are collected from a number of chemical factory sites abandoned after a war, to try to identify what the factories were producing. Eleven substances are measured in each. Principal component analysis (PCA) is used and this identifies scoring systems that indicate the degree of inorganic/organic residues, likelihood of being involved with the textile trade, and likelihood of being involved with weapons manufacture. Groupings of factories are then identified on the basis of these scoring systems.

Cluster analysis (Page 223) – Identifying groupings of similar individuals in a sample or groupings of variables that are closely related

Using the factories example described for principal component analysis, above, cluster analysis is used to identify how individual factories can be grouped in terms of having left similar combinations of chemical residues. The use of cluster analysis to identify groupings of variables that are closely related is also described.

Bibliography

Clarke, G. M. (1994) *Statistics and Experimental Design: An Introduction for Biologists and Biochemists*, 3rd edn. Edward Arnold, London, UK. Quite mathematical in its approach but not as technical as Sokal and Rohlf (1995). It gives some useful detail on experimental design.

Fowler, J., Cohen, L. and Jarvis, P. (1998) *Practical Statistics for Field Biology*, 2nd edn. John Wiley & Sons, Chichester, UK. Covers much of the same ground as this book but is particularly strong on how to deal with datasets made up of counts.

Kenward, M. G. (1987) A method of comparing profiles of repeated measurements. *Applied Statistics*, **36**, 296–308.

Neave, H. R. (1978) *Statistical Tables for Mathematicians, Engineers, Economists and the Behavioural and Management Sciences*. Routledge, London, UK. Contains tables for all of the tests described in this book.

Rowntree, D. (1981) *Statistics Without Tears: An Introduction for Non-mathematicians*. Penguin, London, UK. In some respects this book stops short of giving the amount of detail required to use statistics for practical work but it is an excellent and well-known primer on the basic ideas behind statistics.

Scheirer, C. J., Ray, W. S. and Hare, N. (1976) The analysis of ranked data derived from completely randomized factorial designs. *Biometrics*, **32**, 429–434.

Sokal, R. R. and Rohlf, F. J. (1995) *Biometry*, 3rd edn. W. H. Freeman, New York. A comprehensive text using much technical language, but an excellent reference book for anyone wanting to delve deeper into the subject.

Webster, R. (1997) Regression and functional relations. *European Journal of Soil Science*, **48**, 557–566. See Webster (2001).

Webster, R. (2001) Statistics to support soil research and their presentation. *European Journal of Soil Science*, **52**, 331–340. Two interesting articles aiming to clear up popular misconceptions about the correct use of statistics. They also provide an interesting insight into a journal editor's view of the degree of statistical rigour required for scientific investigations.

Index